808.882
HAR   Harvest of a quiet eye; a selection of
           scientific quotations, by Alan L. Mackay.
           Edited by Maurice Ebison, with a foreword
           by Peter Medawar. Bristol, Institute of
           Physics [c1977]
           192p. illus.

1. Science – Quotations, maxims, etc.   I. Mackay,
Alan Lindsay, ed.   II. Ebison, Maurice, ed.

The Harvest of a Quiet Eye

The harvest of a quiet eye,
That broods and sleeps on his own heart.

From *A Poet's Epitaph*
by William Wordsworth

The Harvest of a Quiet Eye

**A Selection of Scientific Quotations
by Alan L Mackay**

Edited by Maurice Ebison
With a Foreword by Sir Peter Medawar

The Institute of Physics
Bristol and London

This selection and arrangement © 1977 The Institute of Physics

Published by The Institute of Physics
Techno House, Redcliffe Way, Bristol BS1 6NX, and
47 Belgrave Square, London SW1X 8QX

ISBN 0 85498 031 8

British Library Cataloguing in Publication Data
The harvest of a quiet eye: a selection of scientific quotations.
   ISBN 0-85498-031-8
   1. Mackay, Alan Lindsay   2. Ebison, Maurice
   3. Institute of Physics
   808.88′2 PN6084.S/
   Science—Quotations, maxims, etc.

Compiled by Alan Mackay
Edited by Maurice Ebison
Illustrated by John Taylor

Set in 9/11 and 10/12 Times New Roman and 6/7 Univers Medium
Printed in Great Britain by
Graphic Art Services (Brighton) Limited, Burgess Hill, Sussex

Foreword

I was charmed and delighted by this collection of aphorisms and quotations, and hope and expect that many others will be too. Most of them will appeal to scholars generally—very few are for scientists alone, though these few are well chosen (it is fun to read Le Chatelier's theorem as it was propounded by the master himself—93:3).

As is usual with compilations of this kind, some quotations or aphorisms are so persuasive and so well put that one wishes one had said them oneself; others are wrong-headed—even Hardy nods (70:8)—and others still tell us more about their authors than about their subjects: Hilaire Belloc (15:6) cannot have intended to make a public exhibition of himself, but in one short passage he skilfully betrays a total incomprehension of the scientific process, based upon that archaic usage of the word 'experiment', according to which an experiment is an answer to the question: 'I wonder what would happen if?' and again (41:5) the showbiz or cocktail party air of Salvador Dali's comment on DNA makes other quotations seem by comparison more profound than they really are.

The inclusion of a few graffiti was a stroke of genius; my own favourite wall decoration is to be found in the Faculty Club of Rockefeller University, a place where reputations are keenly debated and appraised: a cartoon shows three or four eager scientists discussing the claim to fame of, one presumes, Prometheus: 'Sure, he discovered fire,' the caption runs, 'but what has he done *since?*'

At first sight some of the entries strike the reader as irrelevant to science or scientists, but on closer inspection they will be found either to have a sting or to accord with a train of thought relevant to Dr Mackay's own personal selection of extracts—the labour of many years and clearly a labour of love.

I propose to add one more epigram to this ostensibly irrelevant category: in these days of cost-consciousness, when the funding of research is being administered as if scientific research were a branch of the retail trade, and when 'pure science' and those who practise it are coming under an increasingly cynical scrutiny, it is as well to remember the definition that Oscar Wilde puts into the mouth of Lord Darlington in *Lady Windermere's Fan*: a cynic is 'a man who knows the price of everything and the value of nothing.'

It is strange in any book of quotations not to come upon a great spout of steamy spray from the man Logan Pearsall Smith described as the great Leviathan of English letters, but this is because Dr Johnson did not often bask offshore of the New Atlantis, though when he did so, his observations had a strength, sanity and gravity that will instantly recommend them to the increasing number of younger scientists who are deeply concerned to introduce a moral valuation into science and its applications. In his *Life of Milton*, Johnson chides Milton (and incidentally Cowley) for thinking of an academy in which the scholars should learn astronomy, physics and chemistry in addition to the common run of school subjects. Johnson did not approve of these schemes, for:

'. . . the truth is that the knowledge of external nature and of the sciences which that knowledge requires or includes, is not the great or the frequent business of the human mind. Whether we provide for action or conversation, whether we wish to be useful or pleasing, the first requisite is the religious and moral knowledge of right and wrong; the next is an acquaintance with the history of mankind, and with those examples which may be said to embody truth, and prove by events the reasonableness of opinions. Prudence and justice are virtues, and excellencies, of all times and of all places; we are perpetually moralists, but we are geometricians only by chance. Our intercourse with intellectual nature is necessary; our speculations upon matter are voluntary, and at leisure. Physical knowledge is of such rare emergence, that one man may know another half his life without being able to estimate his skill in hydrostatics or astronomy; but his moral and prudential character immediately appears.'

Sir Peter Medawar
Clinical Research Centre, Harrow

Preface

Scientists have often been reproached for their apparent unfamiliarity with the rest of our cultural inheritance. Inasmuch as science represents one way of dealing with the world, it does tend to separate its practitioners from the rest. Being a scientist resembles membership of a religious order and a scientist usually finds that he has more in common with a colleague on the other side of the world than with his next-door neighbour. But as Shakespeare said of the two cultures: 'Hath not a Jew eyes? Hath not a Jew hands, organs, dimensions, senses, affections, passions, fed with the same food, hurt with the same weapons, subject to the same diseases, healed by the same means, warmed and cooled by the same winter and summer, as a Christian is?' In this case the division was by religion rather than by attitude to science.

Scientists do live in the real world and share in developing and unifying its culture. Many scientists, especially biologists, are able to communicate more widely than to their professional colleagues their sense of wonder at the workings of nature and indeed, for a creative person, the form of the scientific paper is so constrictive that another outlet for his writing is necessary. This collection of quotations is intended to show the wholeness of our culture by demonstrating that scientists contribute to the humanities, and that from Chaucer to Auden, the great humanists have also been concerned with science in all its aspects.

A quotation is a polished prefabricated unit of thought or discourse which has many connotations and associations built in to it. It is thus like the text for a sermon, serving as a point of departure for many lines of thought. Each of us knows many thousands of words and can give, for almost any word, a definition close to that to be found in a dictionary. Yet each one of us has only ever looked up perhaps ten per cent of all the words he knows. We have learnt words by picking them up in their contexts. Each transaction with a word polished it and defined its use and meaning more exactly. Words are coupled into phrases which carry complete thoughts, associations and meanings. We have the subjective feeling, which probably reflects a genuine physical basis, that words are stored in our brains in a vastly ramified network. Extraction of a particular word stirs others and whole phrases and sentences follow. Quotations are, in effect, thoughts embedded in memorable phraseology and polished by use. They are large preformed elements and necessarily combine deep structure (the ideas) with surface structure (the actual words in which the ideas are caught). It is easy to incorporate them, like plug-in circuit boards, into one's own thinking machine.

This is a work of pure plagiarism and I have gleaned items from wherever I came upon them, finding texts sometimes corrupt and full references usually lacking. I apologise for inadequacies and hope that enthusiasts will correct any mistakes and will be stimulated to contribute further entries for later editions. Not all entries in this selection will be familiar to the average scientist, but as quotations are for use, it may help those who write and speak about science to illustrate their material with them and thus some of the less familiar may take root and propagate themselves. Different people will be led along different pathways of thought and some may be stimulated to seek further acquaintance with authors new to them.

In a way the compilation of this book has long been inevitable, because, in 1940, my classics master, S G Squires, required that each pupil in his class should keep a notebook for quotations—I still have mine. So each morning, while the British Empire crumbled, we learnt a new Latin tag and were tested on them once a week. I had doubts of the value of the Classics and even of Shakespeare, but the influence at that formative age had its effect. At the same time the endless exposure to the Bible and the Liturgy of the Church of England provided the essential basis for an informed rationalism and a feeling for the cadences which underlie most of English prose. From the same period, H C Palmer, my senior science master, convinced me that science was interesting and important and set me on a scientific career.

Later, as a prize for the first year examination in Natural Sciences in Cambridge, which was unprecedently postponed because it had been fixed for what turned out to be VE-day, I chose *The Social Function of Science* by J D Bernal. This was my introduction to the works of the marvellous group of encyclopaedists which included Bernal, Waddington, Needham, Haldane, and many other less well known figures. This book was a revelation as to how areas of life which had been disconnected actually fitted together. Later still, after work in industry, I was able to join Bernal's crystallographic laboratory at Birkbeck College, London.

The years in which Bernal was active were immensely stimulating. Besides the revolution in biology all kinds of social, scientific and political movements had their base there and the decrepit buildings which housed the laboratory were an important international and intercultural cross-roads. Alas, in 1963, Bernal suffered a stroke and he passed his last years tragically cut off by his own failing senses. Bernal's example of the excitement and wholeness of life remains an inspiration.

However, it is to my father, whom a book of poetry helped to carry through the Great War, that I would wish to dedicate this selection.

Alan L Mackay
Department of Crystallography, Birkbeck College, London

Introduction

Perhaps the most interesting aspect of editing this selection has been that the resulting book is so very different from the preconceived one that had existed in my imagination. I had expected to be taken on an orderly walk within a restricted field defined by a fairly narrow range of topics from physics and philosophy. What I found instead was a fascinating ramble through an exotic landscape of diverse interests.

In deciding which of the large number of quotations supplied by Dr Mackay should appear in the book every quotation was matched against the following criteria:

(i) Does it stand in its own right in the sense that its meaning can be understood independently of a reader having detailed knowledge of the material from which it is derived?

(ii) Is it likely to encourage at least some readers to want to seek out the source?

(iii) Would it provoke the 'Ah!' reaction, that supreme moment when a flash of unsuspected insight occurs?

(iv) Is it attractive because of real merit or simply because it is already well known?

It is our hope that each of the finally chosen quotations passes at least one of these tests.

The extracts are arranged alphabetically by the name of the author. Occasionally, where it has been helpful to add an explanatory note, this is given within brackets below the quotation itself. Wherever possible the extract is completed by a reference to the source of the quotation. In certain cases the quotation is given first in its original language in italics and is then followed immediately by a translation.

Few things are more irritating than to have only a part of a quotation dodging about in one's mind so elusively that one is unable to pin down either the complete quotation or trace its source. Since no compiler or editor would wish readers to suffer in this way an index of keywords and catch phrases has been provided which should enable a half-remembered quotation to be traced on most occasions. Against each item in the index the number before the colon refers to the page on which the quotation is to be found, while the number after the colon refers to the number of the quotation on that page.

Expressions of gratitude to the professional editorial staff involved in the production of a book are very common in Introductions. This does not mean that they are not heartfelt. Dr Mackay and I know only a little of the expertise and labour that have been involved in securing copyright agreements, in designing an attractive lay-out for the material and in encouraging in a tactful way both compiler and editor. We are sincerely grateful to all in the Institute's Publishing Division who have been involved with this book and in particular to Frances Fawkes, Neville Hankins and Teresa Poole.

Maurice Ebison
The Institute of Physics, London

Russell Lincoln Ackoff 1919–

1   Common sense . . . has the very curious property of being more correct retrospectively than prospectively. It seems to me that one of the principal criteria to be applied to successful science is that its results are almost always obvious retrospectively; unfortunately, they seldom are prospectively. Common sense provides a kind of ultimate validation after science has completed its work; it seldom anticipates what science is going to discover.
Decision-making in National Science Policy 1968 (London: Churchill)

Henry Brooks Adams 1838–1918

2   The future of Thought, and therefore of History, lies in the hands of the physicists, and . . . the future historian must seek his education in the world of mathematical physics. A new generation must be brought up to think by new methods, and if our historical departments in the Universities cannot enter this next phase, the physical departments will have to assume this task alone.
The Degradation of the Democratic Dogma 1919 (New York: Macmillan)

3   All the steam in the world could not, like the Virgin, build Chartres.
[After viewing the Palace of Electricity at the 1900 Trocadero Exposition in Paris]
The Dynamo and the Virgin in *The Education of Henry Adams* 1918 (Boston, Mass: Houghton Mifflin and New York: Heritage Press)

Ahmose the Scribe *ca* 1650 BC

4   In each of 7 houses there are 7 cats; each cat kills 7 mice; each mouse would have eaten 7 hekat of grain. How much grain is saved by the cats?
[Presumably the origin of the *Mother Goose* rhyme: 'As I was going to St Ives . . .']
in *The World of Mathematics* ed J R Newman, 1956 (New York: Simon & Schuster)

5   Ways of investigating Nature and knowing all that exists, every mystery . . . every secret.
[Title of the Rhind Papyrus (on Egyptian mathematics)]
in *The World of Mathematics* ed J R Newman, 1956 (New York: Simon & Schuster)

Mark Akenside 1721–1770

6   Give me to learn each secret cause;
Let number's, figure's motion's laws
Revealed before me stand;
These to great Nature's scene apply,
And round the Globe, and through the sky,
Disclose her working hand.
Hymn to Science in *Works of the English Poets* ed S Johnson, London, 1779, vol 55

Alain [Emile Chartier] 1868–1951

7   There are only two kinds of scholars; those who love ideas and those who hate them.

Alfonso X [King of Castile and Leon] 1221–1284

8   [On having the Ptolemaic system of astronomy explained to him] If the

Lord Almighty had consulted me before embarking upon Creation, I should have recommended something simpler.

Attributed

Luis Alvarez 1911–

1   There is no democracy in physics. We can't say that some second-rate guy has as much right to opinion as Fermi.

in D S Greenberg *The Politics of Pure Science* 1967 (New York: New American Library) © D S Greenberg, 1967

American Philosophical Society

2   . . . . It shall and may be lawful for the said Society by their proper officers, at all times, whether in peace or war, to correspond with learned Societies, as well as individual learned men, of any nation or country . . .

[In its charter of 1780]

Poul Anderson 1926–

3   I have yet to see any problem, however complicated, which, when you looked at it in the right way, did not become still more complicated.

New Scientist 25 September 1969

Anonymous

4   . . . like the statistician who was drowned in a lake of average depth six inches.

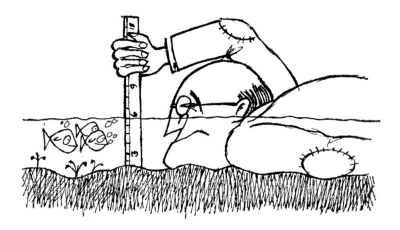

5   'Life is very strange' said Jeremy. 'Compared with what?' replied the spider.

in N Moss *Men who play God* (London: Gollancz)

6   'Tis further from London to Highgate than from Highgate to London.

[An example of a non-commutative metric]
James Howell *Proverbs . . .* 1659

1   Being before the time, the astronomers are to be killed without respite; and being behind the time, they are to be slain without reprieve.
Shu Ching (before 250 BC) in *Nature* 1970 **225** 894

2   I have seen the blacksmith at the mouth of his furnace, his fingers like the skin of a crocodile: he smells worse than the roe of a fish. I have not seen a blacksmith on a commission, a founder who goes on an embassy.
[Written by Egyptian satirist]
Greek Science ed B Farrington, 1963 (London: Pelican/Penguin)

3   An education enables you to earn more than an educator.
in Hans Gaffron *Resistance to Knowledge* 1970 (San Diego, Calif: Salk Institute)

4   Research demands involvement. It cannot be delegated very far.
On Research. A Collection of Quotations 1966 (Cambridge, Mass: A D Little)

5   First baseball umpire: 'Balls and strikes, I call them as I sees them.'
Second umpire: 'Balls and strikes, I call them as they are.'
Third umpire: 'Balls and strikes, they ain't nothing until I call them.'
A Rapoport *Strategy and Conscience* 1969 (New York: Harper & Row)

6   *Gnothi seauton.*
Know thyself.
[From the Temple of Apollo at Delphi]
Pausanias 10.24.1; *Juvenal* 11.27

7   God is not dead: He is alive and well and working on a much less ambitious project.
[Graffito, London, 1975]

8   Half of the secret of resistance to disease is cleanliness; the other half is dirtiness.

9   How many people work in your department?—About a quarter.

10   It is an established rule of the Royal Society . . . never to give their opinion, as a Body, upon any subject, either of Nature or Art, that comes before them.
Advertisement in each issue of the *Philosophical Transactions of the Royal Society* up to the 1950s

11   Laws of Thermodynamics:
1. You cannot win.
2. You cannot break even.
3. You cannot get out of the game.

12   *Magna opera domini exquisita in omnes voluntates eius.*
The works of the Lord are great; sought out of all those that have pleasure therein.
[Over the gateways of the Cavendish Laboratory, Cambridge]

1   A metallurgist is an expert who can look at a platinum blonde and tell whether she is virgin metal or a common ore.

2   Nature requires five,
Custom allows seven,
Idleness takes nine,
And wickedness eleven.
[Hours in bed]
Mother Goose

3   *Nihil in intellectu quod non prius in sensu.*
There is nothing in the mind that has not been previously in the senses.
[See J L Lowes *The Road to Xanadu* for an example of the images which went into Coleridge's poem]

4   Part of the strength of science is that it has tended to attract individuals who love knowledge and the creation of it. Just as important to the integrity of science have been the unwritten rules of the game. These provide recognition and approbation for work which is imaginative and accurate, and apathy or criticism for the trivial or inaccurate . . . . Thus, it is the communication process which is at the core of the vitality and integrity of science . . . . The system of rewards and punishments tends to make honest, vigorous, conscientious, hardworking scholars out of people who have human tendencies of slothfulness and no more rectitude than the law requires.
The Roots of Scientific Integrity. Editorial in *Science* 1963 **139** 3561

5   The Philosophy of Princes is to dive into the secrets of men, leaving the secrets of nature to those that have spare time.
in George Herbert *Jacula Prudentum* 1659

6   Rest in peace. The mistake shall not be repeated.
[Cenotaph in Hiroshima]
Scientific American June 1968

7   Sis, I have found out that there is no Santa Claus, and when I'm a little older, I'm going to look into this stork business, too.
Perspectives in Biology and Medicine 1972, Autumn, p89

8   So, he was a great teacher. But what did he publish?
[From an Israeli academic]

9   This stone commemorates the exploit of William Webb Ellis who, with a fine disregard for the rules of football as played in his time, first took the ball in his arms and ran with it, thus originating the distinctive feature of the Rugby game. AD 1823.
[Perhaps also characteristic of the empirical English school of physics]
Rugby School, England

10   The Turks are not called 'Turks' for nothing.
[This kind of remark leads us to a consideration of names, descriptions and semantics]

1   Twinkle, twinkle little star,
      I don't wonder what you are,
      For by spectroscopic ken
      I know that you are hydrogen.
      in D Bush *Science and English Poetry* 1950 (New York: Oxford UP)

2   What is matter?—Never mind.
      What is mind?—It doesn't matter.

3   When all else fails, read the instructions.

4   We know that the magnet loves the lodestone, but we do not know whether
      the lodestone also loves the magnet or is attracted to it against its will.
      [Arab physicist of the 12th century]
      in D Gabor *Inventing the Future* (London: Secker & Warburg)

Guillaume Apollinaire 1880–1918

5   [Of the Cubists] . . . we who are constantly fighting along the frontiers of
      the infinite and of the future.
      Probably C Grey *Cubist Aesthetic Theories* 1953 (Baltimore, Md: Johns Hopkins Press)

[Saint] Thomas Aquinas *ca* 1225–1274

6   Practical sciences proceed by building up; theoretical sciences by resolving
      into components.
      Commentary on the Ethics I, 3

Arabian Proverb

7   Learning in old age is writing on sand but learning in youth is engraving
      on stone.

François Arago 1786–1853

8   *Connâitre, découvrir, communiquer—telle est la destinée d'un savant.*
      To get to know, to discover, to publish—this is the destiny of a scientist.

Michael A Arbib 1940–

9   In the beginning was the word
      And by the mutations came the gene.
      Towards a Theoretical Biology ed C H Waddington,
      1969 (Edinburgh: Edinburgh UP)

Archimedes 287–212 BC

1   Archimedes to Eratosthenes greeting. . . . certain things first became clear
to me by a mechanical method, although they had to be demonstrated by
geometry afterwards because their investigation by the said method did not
furnish an actual demonstration. But it is of course easier, when we have
previously acquired, by the method, some knowledge of the questions, to
supply the proof than it is to find it without any previous knowledge.
The Method in *The Works of Archimedes* transl T L Heath, 1912 (London: Cambridge UP)

2   *Eureka!*
I have found!
Vitruvius Pollio *De Architectura* ix, 215

3   Give me a firm spot on which to stand, and I will move the earth.
[On the lever]

Aristophanes *ca* 444–*ca* 380 BC

4   First listen, my friend, and then you may shriek and bluster.
Ecclesiazousae 588

5   *Socrates:* Suppose you are arrested for a debt,
We'll say five talents, how will you contrive
To cancel at a stroke both debt and writ?
Strepsiades: I've hit the nail
That does the deed, and so you will confess.
Socrates: Out with it!
Strepsiades: Good chance but you have noted
A pretty toy, a trinket in the shops,
Which being rightly held produceth fire
From things combustible—
Socrates: A burning glass, vulgarly called—
Strepsiades: You are right; 'tis so.
Socrates: Proceed.
Strepsiades: Put the case now your whoreson bailiff comes,
Shows me his writ
—I, standing thus, d'ye mark me,
In the sun's stream, measuring my distance, guide
My focus to a point upon his writ,
And off it goes *in fumo!*
The Clouds transl T Mitchell, 1911 (London: Dent)

Aristotle 384–322 BC

6   If every tool, when ordered, or even of its own accord, could do the work
that befits it, just as the creations of Daedalus moved of themselves . . . .
If the weavers' shuttles were to weave of themselves, then there would be
no need either of apprentices for the master workers or of slaves for the
lords.
Atheniensium Respublica transl F G Kenyon, 1920

1   If one way be better than another, that you may be sure is Nature's way.
      Nichomachean Ethics 1099B, 23

2   If this is a straight line [showing his audience a straight line drawn by a
      ruler], then it necessarily ensues that the sum of the angles of the triangle
      is equal to two right angles; and conversely, if the sum is not equal to two
      right angles, then neither is the triangle rectilinear.
      Physica

3   It is not once nor twice but times without number that the same ideas
      make their appearance in the world.
      On the Heavens in T L Heath *Manual of Greek Mathematics* 1931 (Oxford: Oxford UP)

4   Now that practical skills have developed enough to provide adequately for
      material needs, one of those sciences which are not devoted to utilitarian
      ends [mathematics] has been able to arise in Egypt, the priestly caste there
      having the leisure necessary for disinterested research.
      Metaphysica I–981b

5   Speech is the representation of the mind, and writing is the representation
      of speech.
      De Interpretatione 1

6   The whole is more than the sum of the parts.
      Metaphysica 1045a 10

7   [Quoting Agathon] Chance is beloved of Art, and Art of Chance . . .
      Scientific American August 1969

Neil A Armstrong 1930–

8   One small step for man, one big step for mankind.
      [First words on stepping onto the Moon]
      Nature 1974 **250** 451

9   That's one small step for a man, one giant leap for mankind.
      [Official version]

Thomas Arnold 1795–1842

10   Rather than have Physical Science the principal thing in my son's mind, I
      would rather have him think that the Sun went round the Earth, and that
      the Stars were merely spangles set in a bright blue firmament.

Roger Ascham 1515–1568

11   Mark all mathematical heads which be wholly and only bent on these
      sciences, how solitary they be themselves, how unfit to live with others,
      how unapt to serve the world.
      in E G R Taylor *Mathematical Practitioners of Tudor and Stuart England* 1954 (London: Cambridge
      UP)

[Lord] Eric Ashby 1904–

1   The habit of apprehending a technology in its completeness: this is the
     essence of technological humanism, and this is what we should expect
     education in higher technology to achieve. I believe it could be achieved
     by making specialist studies the core around which are grouped liberal
     studies which are relevant to these specialist studies. But they must be
     relevant; the path to culture should be through a man's specialism, not
     by-passing it . . . . A student who can weave his technology into the fabric
     of society can claim to have a liberal education; a student who cannot
     weave his technology into the fabric of society cannot claim even to be a
     good technologist.
     Technology and the Academics 1958 (London: Macmillan)

John Aubrey 1626–1697

2   . . . 'twas held a strange presumption for a man to attempt improvement of
     any knowledge whatsoever; they thought it not fit to be wiser than their
     fathers and not good manners to be wiser than their neighbours; and a
     sin to search into the ways of nature.
     in Michael Hunter *John Aubrey and the Realm of Learning* 1975 (London: Duckworth)

3   [Of Thomas Hobbes, in 1629] He was 40 years old before he looked on
     geometry; which happened accidentally. Being in a gentleman's library,
     Euclid's *Elements* lay open, and 'twas the 47 *El. libri* I [Pythagoras'
     Theorem]. He read the proposition. 'By God,' sayd he. (He would now
     and then swear, by way of emphasis.) 'By God,' sayd he, 'this is im-
     possible:' So he reads the demonstration of it, which referred him back to
     such a proposition; which proposition he read. That referred him back to
     another, which he also read. *Et sic deinceps*, that at last he was demon-
     stratively convinced of that trueth. This made him in love with geometry.
     Brief Lives ed O L Dick, 1960 (Oxford: Oxford UP)

Wystan Hugh Auden 1907–1973

4   But he would have us remember most of all
     To be enthusiastic over the night
     Not only for the sense of wonder
     It alone has to offer, but also
     Because it needs our love: for with sad eyes
     Its delectable creatures look up and beg
     Us dumbly to ask them to follow;
     They are exiles who long for the future.
     In Memory of Sigmund Freud 1951 (New Haven, Conn: Yale UP)

5   How happy the lot of the mathematician. He is judged solely by his peers,
     and the standard is so high that no colleague or rival can ever win a repu-
     tation he does not deserve.
     The Dyer's Hand 1948 (London: Faber & Faber)

6   Love requires an Object,
     But this varies so much,

Almost, I imagine,
Anything will do:
When I was a child, I
Loved a pumping-engine,
Thought it every bit as
Beautiful as you.
Heavy Date in *Collected Shorter Poems, 1927–1957* 1966 (London: Faber & Faber)

1   Of course, Behaviourism 'works'. So does torture. Give me a no-nonsense, down-to-earth behaviourist, a few drugs, and simple electrical appliances, and in six months I will have him reciting the Athanasian Creed in public.
A Certain World 1970 (London: Faber & Faber)

2   Those who will not reason
Perish in the act:
Those who will not act
Perish for that reason.
in *Shorts* in *Collected Shorter Poems, 1927–1957* 1966 (London: Faber & Faber)

3   Thou shalt not answer questionnaires
Or quizzes upon world affairs,
Nor with compliance
Take any test.
Thou shalt not sit with statisticians nor commit
A social science.
Under which lyre in *Collected Poems of W H Auden* (London: Faber & Faber)

4   To the man-in-the-street, who, I'm sorry to say
Is a keen observer of life,
The word *intellectual* suggests right away
A man who's untrue to his wife.
Note on Intellectuals in *Shorts* in *Collected Shorter Poems, 1927–1957* 1966 (London: Faber & Faber)

5   Tomorrow, perhaps, the future: the research on fatigue
And the movements of packers; the gradual exploring of all the
Octaves of radiation; . . .
Spain 1937 in *Collected Poems of W H Auden* (London: Faber & Faber)

6   When I find myself in the company of scientists, I feel like a shabby curate who has strayed by mistake into a drawing room full of dukes.
The Dyer's Hand and Other Essays 1962 (New York: Random House)

Pierre Auger 1899–

7   Holophrase—multisyllabic words designating precisely a very complex situation but entirely *sui generis* and not resolvable; that is, containing neither roots nor affixes used elsewhere. The best known example is *mamihlapinatapai* which in Tierra del Fuego designates the situation in which two persons look at each other, each hoping to see the other under-

take an action which both wish for without being disposed to take the
initiative.

The Regime of Castes in the Populations of Ideas in *Diogenes* 1958 **22** 42

Saint Augustine 354–430

1   *Angelus non potest peccare; homo potest non peccare.*
An angel cannot sin; a man can choose not to sin.

2   The good Christian should beware of mathematicians [astrologers], and all
those who make empty prophecies. The danger already exists that the
mathematicians have made a covenant with the devil to darken the spirit
and to confine man in the bonds of Hell.

De Genesi ad Litteram book II, xviii, 37

3   If I am given a sign [formula], and I am ignorant of its meaning, it cannot
teach me anything, but if I already know it what does the sign teach me?

De Magistro ch X, 23

4   *Nisi credideritis, non intelligitis.*
If you don't believe it, you won't understand it.

De Libero Arbitrio

Marcus Aurelius Antoninus 121–180

5   Everything that happens, happens as it should, and if you observe carefully,
you will find this to be so.

Meditations IV, 10

6   That which is not good for the bee-hive, cannot be good for the bees.

[Like General Motors and the USA]
Meditations IV, 49

7   The world is either the effect of cause or of chance. If the latter, it is a
world for all that, that is to say, it is a regular and beautiful structure.

Meditations IV, 22

Avicenna [Ibn Sina] 980–1037

8   *Non turpe est medico, cum de rebus veneris loquitur, de delectatione mulieris
coeuntis: quoniam sunt ex causis, quibus pervenitur ad generationem.*
Writing about erotics is a perfectly respectable function of medicine, and
about the way to make the woman enjoy sex; these are an important part
of reproductive physiology.

in Alex Comfort *Sex in Society* 1963 (London: Duckworth)

Azerbaijani Proverb

9   Speak not about what you have read, but about what you have understood.

Azerbaijani Wall Poster

10   Without sanitary culture, there would be no culture at all.

[Seen on a banner outside the Academy of Sciences in Baku, 1962]

Charles Babbage 1792–1871

1   Errors using inadequate data are much less than those using no data at all.

2   Remember that accumulated knowledge, like accumulated capital, increases at compound interest: but it differs from the accumulation of capital in this; that the increase of knowledge produces a more rapid rate of progress, whilst the accumulation of capital leads to a lower rate of interest. Capital thus checks its own accumulation: knowledge thus accelerates its own advance. Each generation, therefore, to deserve comparison with its predecessor, is bound to add much more largely to the common stock than that which it immediately succeeds.
The Exposition of 1851 1851 (London: Murray)

3   That the master manufacturer, by dividing the work to be executed into different processes, each requiring different degrees of skill and strength, can purchase exactly that precise quantity of both which is necessary for each process; whereas, if the whole work were executed by one workman, that person must possess sufficient skill to perform the most difficult, and sufficient strength to execute the most laborious, of the various operations . . . . When (from the peculiar nature of the produce of each manufactory) the number of processes into which it is most advantageous to divide it is ascertained, as well as the number of individuals to be employed, then all other manufactories which do not employ a direct multiple of this number, will produce the article at greater cost. This principle ought always to be kept in view in great establishments, although it is quite impossible, even with the best system of the division of labour, to carry it rigidly into execution . . . .
On the Economy of Machinery and Manufacturers London, 1832

4   The whole of the developments and operations of analysis are now capable of being executed by machinery . . . . As soon as an Analytical Engine exists, it will necessarily guide the future course of science.
Passages from the Life of a Philosopher 1864 (London: Longman)

Gaston Bachelard 1884–1962

5   *L'observation scientifique est toujours une observation polémique; elle confirme ou infirme une thèse antérieure; un schéma préalable, un plan d'observation; elle montre en démontrant; elle hierarchise les apparences; elle trancende l'immédiat; elle reconstruit le réel aprés avoir reconstruit ses schémas.*
A scientific observation is always a committed observation; it confirms or denies one's preconceptions; one's first ideas, one's plan of observation; it shows by demonstration; it structures the phenomena; it transcends what is close at hand; it reconstructs the real after having reconstructed its representation.
La nouvel ésprit scientifique Introduction

Francis Bacon [Lord Verulam] 1561–1626

6   The End of our Foundation is the knowledge of **Causes**, and the secret

motions of things; and the enlarging of the bounds of Human Empire, to the effecting of all things possible.
New Atlantis 1626

1   For man being the minister and interpreter of nature, acts and understands so far as he has observed of the order, the works and mind of nature, and can proceed no further; for no power is able to loose or break the chain of causes, nor is nature to be conquered but by submission: whence those twin intentions, human knowledge and human power, are really coincident; and the greatest hindrance to works is the ignorance of causes.
The Great Instauration Preface

2   He that will not apply new remedies must expect new evils; for time is the greatest innovator.
On Innovations Essays

3   The human Intellect, in those things which have once pleased it (either because they are generally received and believed, or because they suit the taste), brings everything else to support and agree with them; and though the weight and number of contradictory instances be superior, still it either overlooks or despises them, or gets rid of them by creating distinctions, not without great and injurious prejudice, that the authority of these previous conclusions may be maintained inviolate. And so he made a good answer, who, when he was shown, hung up in a temple, the votive tablets of those who had fulfilled their vows after escaping from shipwreck, and was pressed with the question, 'Did he not then recognize the will of the gods?' asked, in his turn, 'But where are the pictures of those who have perished, notwithstanding their vows?' The same holds true of almost every superstition—as astrology, dreams, omens, judgments, and the like—wherein men, pleased with such vanities, attend to those events which are fulfilments; but neglect and pass over the instances where they fail (though this is much more frequently the case).
Novum Organum 1620

4   It is well to observe the force and virtue and consequence of discoveries, and these are to be seen nowhere more conspicuously than in printing, gunpowder and the magnet. For these three have changed the whole face and state of things throughout the world, . . . in so much that no empire, no sect, no star seems to have exerted greater power and influence in human affairs than these mechanical discoveries.
Novum Organum 1620

5   *Nam et ipsa scientia potestas est.*
For knowledge itself is power.
Religious Meditations, Of Heresies

6   *Natura non nisi parendo vincitur.*
Nature, to be commanded, must be obeyed.
Novum Organum 1620

1   Reading maketh a full man; conference a ready man; and writing an exact man.
[Bacon maketh a fat man—graffito]
Of Studies Essay 50

2   Truth comes out of error more readily than out of confusion.
Novum Organum 1620

3   Histories make men wise; poets, witty; the mathematics subtile; natural philosophy, deep; moral, grave; logic and rhetoric, able to contend.
Of Studies Essay 50

4   The ill and unfit choice of words wonderfully obstructs the understanding.
Novum Organum 1620

Roger Bacon 1220–1292

5   *Et harum scientarum porta et clavis est Mathematica.*
Mathematics is the door and the key to the sciences.
Opus Majus transl Robert Belle Burke, 1928 (Philadelphia, Pa: University of Pennsylvania Press)

6   For the things of this world cannot be made known without a knowledge of mathematics. For this is an assured fact in regard to celestial things, since two important sciences of mathematics treat of them, namely theoretical astrology and practical astrology. The first . . . gives us definite information as to the number of the heavens and of the stars, whose size can be comprehended by means of instruments, and the shapes of all and their magnitudes and distances from the earth, and thicknesses and number, and greatness and smallness . . . . It likewise treats of the size and shape of the habitable earth . . . . All this information is secured by means of instruments suitable for these purposes, and by tables and by canons . . . . For everything works through innate forces shown by lines, angles and figures.
Opus Majus transl Robert Belle Burke, 1928 (Philadelphia, Pa: University of Pennsylvania Press)

Walter Bagehot 1826–1877

7   You may talk of the tyranny of Nero and Tiberias, but the real tyranny is the tyranny of your next-door neighbour. What espionage of despotism comes to your door so effectively as the eye of the man who lives at your door? Public opinion is a permeating influence. It requires us to think other men's thoughts, to speak other men's words, to follow other men's habits.
The Works of Walter Bagehot 1889 (Hartford, Conn: Traveler's Insurance Co)

8   One of the greatest pains to human nature is the pain of a new idea.
Physics and Politics in *Collected Works* ed N A F St J Stevas, 1965 (London: Economist)

Arthur [Earl of] Balfour 1848–1930

9   [Brother-in-law of Lord Rayleigh] . . . science was the great instrument of social change, all the greater because its object is not change but knowledge and its silent appropriation of this dominant function, amid the din of

political and religious strife, is the most vital of all the revolutions which have marked the development of modern civilisation.

Decadence 1908 (London: Cambridge UP)

Walter William Rouse Ball 1850–1925

1   The manner of de Moivre's death has a certain interest for psychologists. Shortly before it, he declared that it was necessary for him to sleep some ten minutes or a quarter of an hour longer each day than the preceding one: the day after he had thus reached a total of something over twenty-three hours he slept up to the limit of twenty-four hours, and then died in his sleep.

[Abraham de Moivre 1667–1754]
History of Mathematics 1911 (London: Macmillan)

William Bateson 1861–1926

2   I would trust Shakespeare, but I would not trust a committee of Shakespeares.

[The geneticist]
in J K Brierley *Biology and the Social Crisis* 1967 (London: Heinemann Education)

Edmone-Pierre Chanvot de Beauchêne 1748–1824

3   Science seldom renders men amiable; women, never.
Maximes, réflexions et pensées diverses

Stafford Beer 1926–

4   *Absolutum obsoletum*—if it works it's out of date.
Brain of the Firm 1972 (London: Penguin)

1   Man: 'Hello, my boy. And what is your dog's name?'
     Boy: 'I don't know. We call him Rover.'
     New Scientist 3 October 1974

2   Our institutions are failing because they are disobeying laws of effective
     organisation which their administrators do not know about, to which indeed
     their cultural mind is closed, because they contend that there exists and can
     exist, no science competent to discover those laws.
     Designing Freedom 1974 (Chichester: Wiley)

Belgian Notice

3   *Ne parler pas au wattman.*
     Do not talk to the tram-driver.
     [Thus immortalising James Watt]

Vissarion Grigorievich Belinskii 1811–1848

4   In science one must search for ideas. If there are no ideas then there is no
     science. A knowledge of facts is only valuable in so far as facts conceal
     ideas: facts without ideas clutter up the mind and the memory.
     Sobranie sochinenii 1948 (Moscow: OGIZ)

Eric Temple Bell 1883–1960

5   The cowboys have a way of trussing up a steer or a pugnacious bronco
     which fixes the brute so that it can neither move nor think. This is the
     hog-tie, and it is what Euclid did to geometry.

Hilaire Belloc 1870–1953

6   Anyone of common mental and physical health can practise scientific
     research . . . . Anyone can try by patient experiment what happens if this
     or that substance be mixed in this or that proportion with some other
     under this or that condition. Anyone can vary the experiment in any
     number of ways. He that hits in this fashion on something novel and of
     use will have fame . . . . The fame will be the product of luck and industry.
     It will not be the product of special talent.
     Essays of a Catholic Layman in England 1931 (London: Sheed & Ward)

Nikolai Vassilevich Belov 1891–

7   [Obituary of J D Bernal] . . . like a true Irishman, his last enthusiasm was
     for the laws of lawlessness.
     [Bernal developed a geometrical theory of liquids]
     Soviet Physics–Crystallography 1972 **17** 208–9

[Enoch] Arnold Bennett 1867–1931

8   [Of *Nature*] The writing of it is considerably inferior to the matter of it . . . .
     I regard *Nature* as perhaps the most important weekly printed in English,
     far more important than any political weekly.
     Evening Standard 20 November 1930

Henry Albert Bent 1926–

1   . . . hell must be isothermal, for otherwise the resident engineers and physical chemists (of which there must be some) could set up a heat engine to run a refrigerator to cool off a portion of their surroundings to any desired temperature.
The Second Law 1965 (New York: Oxford UP)

2   The important point is not the bigness of Avogadro's number [6×10^{23} atoms/g atom] but the bigness of Avogadro.
[Avogadro consisted of some 10^{27} atoms]
The Second Law 1965 (New York: Oxford UP)

Jeremy Bentham 1748–1832

3   O Logic: born gatekeeper to the Temple of Science, victim of capricious destiny: doomed hitherto to be the drudge of pedants: come to the aid of thy master, Legislation.
Works ed J Browning, 1838–1843

Edmund Clerihew Bentley 1875–1956

4   Sir Humphrey Davy
Detested gravy.
He lived in the odium
Of having discovered Sodium.
Biography for Beginners 1925 (London: Werner Laurie)

Nicolas Berdiaeff 1874–1948

5   *Les utopies apparaissent comme bien plus réalisables qu'on ne le croyait autrefois. Et nous nous trouvons actuellement devant une question bien autrement angoissante: Comment éviter leur réalisation définitive?*
Utopias now appear much more realizable than one used to think. We are now faced with a very different new worry: How to prevent their realization.
in Aldous Huxley *Brave New World* 1932 (London: Chatto & Windus)

Henri Bergson 1859–1941

6   *L'univers . . . est une machine à faire des dieux.*
The universe . . . is a device for making deities.
Les deux sources de la morale et de la religion 1932 (Paris: Presses Universitaires de France)

John Desmond Bernal 1901–1971

7   All that glisters may not be gold, but at least it contains free electrons.
[But consider the Golden Scarab Beetle which has a metallic lustre without metal]
Lecture at Birkbeck College, University of London, 1960

8   But if capitalism had built up science as a productive force, the very character of the new mode of production was serving to make capitalism itself unnecessary.
Marx and Science 1952 (London: Lawrence & Wishart)

1   The greater the man the more he is soaked in the atmosphere of his time; only thus can he get a wide enough grasp of it to be able to change substantially the pattern of knowledge and action.
Science in History 1954 (London: Watts)

2   In fact, we will have to give up taking things for granted, even the apparently simple things. We have to learn to understand nature and not merely to observe it and endure what it imposes on us. Stupidity, from being an amiable individual defect, has become a social crime.
The Origin of Life 1967 (London: Weidenfeld & Nicolson)

3   It is characteristic of science that the full explanations are often seized in their essence by the percipient scientist long in advance of any possible proof.
The Origin of Life 1967 (London: Weidenfeld & Nicolson)

4   Life is a partial, continuous, progressive, multiform and conditionally interactive self-realisation of the potentialities of atomic electron states.
The Origin of Life 1967 (London: Weidenfeld & Nicolson)

Claude Bernard 1813–1878

5   *La science n'admet pas les exceptions; sans cela il n'y aurait aucun déterminisme dans la science, ou plutôt il n'y aurait plus de science.*
Science allows no exceptions; without this there would be no determinism in science, or rather, there would be no science at all.
Leçons de pathologie expérimentale 1871

6   A modern poet has characterised the personality of art and the impersonality of science as follows: art is I; science is we.
Introduction à l'étude de la médecine expérimentale 1865, I, 2.4, line 1742

Edward Bernard 17th Century

7   Books and experiments do well together, but separately they betray an imperfection, for the illiterate is anticipated unwillingly by the labours of the ancients, and the man of authors deceived by story instead of science (1671).
in S J Rigaud *Correspondence of Scientific Men of the Seventeenth Century* Oxford, 1841, vol 1

Jacques Bernouilli 1654–1705

8   We define the art of conjecture, or stochastic art, as the art of evaluating as exactly as possible the probabilities of things, so that in our judgments and actions we can always base ourselves on what has been found to be the best, the most appropriate, the most certain, the best advised; this is the only object of the wisdom of the philosopher and the prudence of the statesman.
Ars Conjectandi Basel, 1713. Transl in B de Jouvenel *The Art of Conjecture* 1967 (London: Weidenfeld & Nicolson)

Jöns Jacob von Berzelius [Baron Berzelius] 1779–1848

9   [On moving from collecting to experimental work] Immediately with the

first participation in them I was seized with a feeling never previously experienced; I was irrevocably gripped by this method of pursuing knowledge. I must needs repeat for myself the experiments I had seen him [Ekmark] perform, and although I was unable to buy any instruments, I improvised apparatus, with his help, which I myself could make . . . . [When he first collected oxygen in a laboratory exercise] I . . . have seldom experienced a moment of such pure and deep happiness as when the glowing stick which was thrust into it lighted up and illuminated with unaccustomed brilliancy my windowless laboratory.
Autobiographical Notes

Bhartrhari 5th or 6th Century

1   In this vain fleeting universe, a man
Of wisdom has two courses: first, he can
Direct his time to pray, to save his soul,
And wallow in religion's nectar-bowl;
But, if he cannot, it is surely best
To touch and hold a lovely woman's breast,
And to caress her warm round hips, and thighs,
And to possess that which between them lies.
[D D Kosambi (1907–1966) the editor of the Sanskrit text, was an Indian mathematician of wide learning]
in D D Kosambi *Satakatrayadi-subhasitasamgraha: The Epigrams attributed to Bhartrhari*

Bhaskara [Acharya] 1114–*ca* 1185

2   Beautiful and dear Lilavati, whose eyes are like a fawn's . . .? How many are the variations of form of the god Chambhu by the exchange of his ten attributes held reciprocally in his several hands: namely the rope, the elephant hook, the serpent, the tabor, the skull, the trident, the bedstead, the dagger, the arrow and the bow . . .?
Sidd'hanta-siromani chapter in *Lilavati* (*ca* 1150) transl H T Colebrook, 1817 (London: Murray)
II. 16 and XIII. 269

3   A particle of tuition conveys science to a comprehensive mind; and having reached it, expands of its own impulse. As oil poured upon water, as a secret entrusted to the vile, as alms bestowed upon the worthy, however little, so does science infused into a wise mind spread by intrinsic force.
Conclusion of *Vija-Ganita* chapter in *Lilavati* (*ca* 1150) transl H T Colebrook, 1817 (London: Murray)

The Bible

4   Oh . . . my desire is . . . that mine adversary had written a book.
Job 31: 35

5   Where wast thou when I laid the foundations of the earth? Declare if thou hast understanding.
Who hath laid the measures thereof, if thou knowest? Or who hath stretched the line upon it?
Hast thou entered into the springs of the sea? Or hast thou walked in the search of the depth?

Have the gates of death been opened unto thee? Or hast thou seen the doors of the shadow of death?
Canst thou bind the sweet influences of Pleiades, or loose the bands of Orion?
Canst thou lift up thy voice to the clouds, that abundance of waters may cover thee? Canst thou send lightnings, that they may go, and say unto thee, Here we are? Who hath put wisdom in the inward parts? Or who hath given understanding to the heart? . . . who can stay the bottles of heaven?
Job 38: 4–5; 16–17; 31; 34–37

1   Evil devices are an abomination to the Lord: but pleasant words are pure.
Proverbs 15: 26

2   Where there is no vision the people perish: . . .
Proverbs 29: 18

3   . . . he that increaseth knowledge, increaseth sorrow.
Ecclesiastes 1: 18

4   And furthermore, my son, be admonished: of making many books there is no end; and much study is a weariness of the flesh.
Ecclesiastes 12: 12

5   Who was it who measured the water of the sea in the hollow of his hand and calculated the dimensions of the heavens,
gauged the whole earth to the bushel,
weighed the mountains in scales,
the hills in a balance?
Isaiah 40: 12 in *The Jerusalem Bible*

6   And the king spake unto Ashpenaz the master of his eunuchs, that he should bring certain of the Children of Israel, and of the king's seed, and of the princes';
Children in whom was no blemish, but well favoured, and skilful in all wisdom, and cunning in knowledge, and understanding science, and such as had ability in them to stand in the king's palace, and whom they might teach the learning and the tongue of the Chaldeans.
[Those chosen were Belteshazzar, Shadrach, Meshach and Abed-nego, whose history is discussed further on in the chapter. This is the only mention of science in the Old Testament]
Daniel 1: 3–4

7   . . . MENE, MENE, TEKEL, UPHARSIN.
[Said to be Aramaic for: numbered, numbered, weighed, divided]
Daniel 5: 25

8   And should not I spare Nineveh, that great city, wherein are more than sixscore thousand persons that cannot discern between their right hand and their left hand; and also much cattle?
Jonah 4: 11

1   For unto every one that hath shall be given, and he shall have abundance: but from him that hath not shall be taken away even that which he hath.
[Matthew Principle of Scientific Publication enunciated by R K Merton]
Matthew 25: 29

2   For which of you, intending to build a tower, sitteth not down first, and counteth the cost, whether he have sufficient to finish it?
Luke 14: 28

3   And ye shall know the truth, and the truth shall make you free.
[Inscribed on the wall of the main lobby at the CIA Headquarters, Langley, Virginia, USA]
John 8: 32

4   For all the Athenians and strangers which were there spent their time in nothing else, but either to tell, or to hear some new thing.
Acts 17: 21

5   O Timothy, keep that which is committed to thy trust, avoiding profane and vain babblings, and oppositions of science falsely so called.
[This is the only mention of science in the New Testament—in Greek, *gnosis*]
1 Timothy 6: 20

6   And I went unto the angel, saying unto him that he should give me the little book. And he saith unto me, Take it and eat it up; and it shall make thy belly bitter, but in thy mouth it shall be as sweet as honey.
Revelations 10: 9

7   . . . . But by measure and number and weight, thou didst order all things . . . .
Wisdom of Solomon (Apocrypha) 11: 20

8   He that sinneth before his Maker, Let him fall into the hands of the physician.
Ecclesiasticus (Apocrypha) 38: 15

Al-Biruni 973–1048

9   Once a sage was asked why scholars always flock to the doors of the rich, whilst the rich are not inclined to call at the doors of scholars. 'The scholars,' he answered, 'are well aware of the use of money, but the rich are ignorant of the nobility of science.'

10   [On the science and culture of the Hindus] I can only compare their astronomical and mathematical literature . . . to a mixture of pearl shells and sour dates, or of costly crystals and common pebbles. Both kinds of things are equal in their eyes, since they cannot rise themselves to the methods of strictly scientific deduction.
Hindustan transl C E Sachau, London, 1888

Otto von Bismarck 1815–1898

11   *Die Politik ist keine exakte Wissenschaft.*

Politics is not an exact science.
Speech, Prussian Chamber, 18 December 1863

Patrick Maynard Stuart Blackett 1897–1974

1   A first-rate laboratory is one in which mediocre scientists can produce outstanding work.
by M G K Menon in his Commemoration Lecture on H J Bhabha, 1967 *The Royal Institution*

2   The scientist can encourage numerical thinking on operational matters and so help avoid running the war by gusts of emotion.
Operational Research in the RAF (London: The Air Ministry, HMSO)

William Blake 1757–1827

3   [On industrialisation] Hampstead, Highgate, Finchley, Muswell Hill rage loud
Before Bromion's iron Tongs and glowing Poker reddening fierce; . . .
Jerusalem plate 16, 1.1–2

4   The Atoms of Democritus
And Newton's Particles of Light
Are sands upon the Red Sea shore
Where Israel's tents do shine so bright.
Mock on, Mock on: Voltaire, Rousseau ca 1800

5   Consider Sexual Organisation and hide thee in the dust.
Jerusalem plate 34, ch 2

6   He who would do good to another must do it in Minute Particulars:
General Good is the plea of the scoundrel, hypocrite and flatterer,
For Art and Science cannot exist but in minutely organised particulars.
Jerusalem plate 55, 60–1

7   I must Create a System, or be enslaved by another Man's;
I will not Reason and Compare; My business is to Create.
Jerusalem plate 10, 20

8   I was in a Printing-house in Hell, and saw the method in which knowledge is transmitted from generation to generation.
The Marriage of Heaven and Hell

9   To teach doubt and Experiment
Certainly was not what Christ meant.
Blake Complete Writings

Niels Henrik David Bohr 1885–1962

10   . . . two sorts of truth: trivialities, where opposites are obviously absurd, and profound truths, recognised by the fact that the opposite is also a profound truth.
My Father in *Niels Bohr: his life and work* . . . ed S Rozental, 1967 (New York: Wiley)

1   An expert is a man who has made all the mistakes, which can be made, in a very narrow field.
Edward Teller, 10 November 1972, US Embassy

2   When it comes to atoms, language can be used only as in poetry. The poet too, is not nearly so concerned with describing facts as with creating images.
in J Bronowski *The Ascent of Man* 1975 (London: BBC)

Ludwig Boltzmann 1844–1906

3    $S = k \log \Omega$
[Carved above the name of Ludwig Boltzmann on his tombstone in the Zentralfriedhof in Vienna]

Wolfgang Bolyai 1775–1856

4   Detest it just as much as lewd intercourse; it can deprive you of all your leisure, your health, your rest, and the whole happiness of your life.
[Letter to his son János, warning him to give up his attempts to prove the Euclidean postulate on parallels]

Étienne Bonnot [Abbé de Condillac] 1714–1780

5   *Voulez vous apprendre les sciences avec facilité? Commencez par apprendre votre language.*
Do you wish to learn science easily? Then begin by learning your own language.
Essai sur l'origine des connaissances humaines

Andrew Donald Booth 1918–

6   Every system is its own best analogue.
[Over tea at Birkbeck College, University of London, *ca* 1955]

Max Born 1882–1970

7   I am now convinced that theoretical physics is actual philosophy.
Autobiography

Roger Joseph Boscovich 1711–1780

8   *Homo hominem arreptum a Tellure, et ubicumque exigua impulsum vi vel uno etiam oris flatu impetitum, ab hominum omnium commercio in infinitum expelleret, nunquam per totam aeternitatem rediturum.*
Were it not for gravity one man might hurl another by a puff of his breath into the depths of space, beyond recall for all eternity.
Theoria Philosophiae Naturalis Vienna 1758, par 552. English transl J M Child, 1922 (La Salle, Ill: Open Court)

James Boswell 1740–1795

9   When Dr Johnson felt, or fancied he felt, his fancy disordered, his constant recurrence was to the study of arithmetic.
Life of Johnson Harper's edn, 1871. vol 2

Gordon Bottomley 1874–1948

10   Your worship is your furnaces,

Which, like old idols, lost obscenes,
Have molten bowels; your vision is
Machines for making more machines.
To Iron Founders and Others in *Poems of Thirty Years* 1925 (London: Constable)

Pierre Boulez 1925–

1   Music cannot move forward without science.
The Observer 27 July 1975

Matthew Boulton 1711–1780

2   I am selling what the whole world wants; power.
[Letter to Catherine the Great of Russia offering steam engines for sale]
in J D Bernal *Science in History* (Cambridge, Mass: The MIT Press)

Nicholas Bourbaki (pseudonym)

3   Structures are the weapons of the mathematician.
[Collective pseudonym of the Nancy school of mathematics. See *Scientific American* May 1957]

Francis Herbert Bradley 1846–1924

4   Metaphysics is the finding of bad reasons for what we believe upon instinct.
Appearance and Reality Preface

[Sir] William Lawrence Bragg 1890–1971

5   The electron is not as simple as it looks.
Recounted by Sir George Paget Thompson at the *Electron Diffraction Conference* Imperial College,
University of London, 1967

6   The important thing in science is not so much to obtain new facts as to
discover new ways of thinking about them.
in A Koestler and J R Smithies *Beyond Reductionism* 1968 (London: Hutchinson)

Tycho Brahe 1546–1601

7   And when statesmen or others worry him [the scientist] too much, then he
should leave with his possessions. With a firm and steadfast mind one
should hold under all conditions, that everywhere the earth is below and
the sky above, and to the energetic man, every region is his fatherland.
[The 'brain drain' has existed as long as science]
Denmark, 1597

8   Now it is quite clear to me that there are no solid spheres in the heavens,
and those that have been devised by the authors to save the appearances,
exist only in the imagination, for the purpose of permitting the mind to
conceive the motion which the heavenly bodies trace in their courses.
['Saving the appearances' is the old expression for fitting the theory to the facts]

Georges Braque 1882–1963

9   *L'art est fait pour troubler. La science rassure.*
Art upsets, science reassures.
Pensées sur l'Art (Paris: Gallimard)

1   *La vérité existe. On n'invente que la mensonge.*
The truth exists—only fictions are invented.
Pensées sur l'Art (Paris: Gallimard)

Bertolt Brecht 1898–1956

2   *Aus den Bücherhallen*
Treten die Schlächter.
Die Kinder an sich drückend
Stehen die Mütter und durchforschen entgeistert
Den Himmel nach den Erfindungen der Gelehrten.
Out of the libraries
Come the slaughterers.
Pressing their children to them,
Mothers stand shocked, scanning the skies for the inventions of the
professors.
1940 Werkausgabe, Suhrkamp, Band 9

3   *Und was nützt freie Forschung ohne freie Zeit zu forschen?*
What good is freedom to research without free time to do it in?
Leben des Galilei 1958 (Berlin: Aufbau-Verlag) scene 1

[Sir] David Brewster 1781–1868

4   And why does England thus persecute the votaries of her science? Why
does she depress them to the level of her hewers of wood and her drawers
of water? It is because science flatters no courtier, mingles in no political
strife . . . . Can we behold unmoved the science of England, the vital
principle of her arts, struggling for existence, the meek and unarmed victim
of political strife?
Quarterly Review 1830 **43** 320, 323–4 (reviewing Babbage's book *Reflexions on the Decline of Science in England*)

5   The infant [Newton] . . . ushered into the world was of such diminutive
size, that, as his mother afterwards expressed it to Newton himself, he
might have been put into a quart-mug . . . .
Memoirs of Newton 1855

Robert Bridges 1844–1930

6   . . . we only think to find
A structure of blind atoms to their habits enslaved.
The Testament of Beauty 1930

7   Now will the Orientals make hither in return
Outlandish pilgrimage; their wise acres have seen
The electric light; in the West, and come to worship.
The Testament of Beauty 1930

Anthelme Brillat-Savarin 1755–1826

8   *La destinée des nations dépend de la manière dont elles se nourrissent.*
The destiny of countries depends on the way they feed themselves.
Physiologie du Goût 1825

Louis Victor de Broglie 1892–

1   Two seemingly incompatible conceptions can each represent an aspect of the truth . . . . They may serve in turn to represent the facts without ever entering into direct conflict.
Dialectica I, 326

Jacob Bronowski 1908–1974

2   . . . no science is immune to the infection of politics and the corruption of power . . . . The time has come to consider how we might bring about a separation, as complete as possible, between Science and Government in all countries. I call this the disestablishment of science, in the same sense in which the churches have been disestablished and have become independent of the state.
Encounter July 1971

3   By the worldly standards of public life, all scholars in their work are of course oddly virtuous. They do not make wild claims, they do not cheat, they do not try to persuade at any cost, they appeal neither to prejudice nor to authority, they are often frank about their ignorance, their disputes are fairly decorous, they do not confuse what is being argued with race, politics, sex or age, they listen patiently to the young and to the old who both know everything. These are the general virtues of scholarship, and they are peculiarly the virtues of science.
Science and Human Values 1956 (London: Hutchinson)

4   The hand is the cutting edge of the mind.
The Ascent of Man 1975 (London: BBC)

5   It is important that students bring a certain ragamuffin, barefoot irreverence to their studies; they are not here to worship what is known, but to question it.
The Ascent of Man 1975 (London: BBC)

6   Man masters nature not by force but by understanding. That is why science has succeeded where magic failed: because it has looked for no spell to cast on nature.
Science and Human Values 1956 (London: Hutchinson)

7   Science has nothing to be ashamed of, even in the ruins of Nagasaki.
Science and Human Values (New York: Julian Messner (Simon & Schuster))

Jacob Bronowski and **Bruce Mazlish** 1908–1974 and 1923–

8   Every thoughtful man who hopes for the creation of a contemporary culture knows that this hinges on one central problem: to find a coherent relation between science and the humanities.
The Western Intellectual Tradition 1960 (London: Hutchinson)

Rupert Brooke 1887–1915

9   But somewhere, beyond Space and Time

Is wetter water, slimier slime.
And there (they trust) there swimmeth One
Who swam ere rivers were begun,
Immense, of fishy form and mind,
Squamous, omnipotent, and kind.
Heaven

1   For Cambridge people rarely smile,
Being urban, squat, and packed with guile; . . .
The Old Vicarage, Grantchester written Berlin, 1912

George Spencer Brown 1923–

2   To arrive at the simplest truth, as Newton knew and practised, requires
years of contemplation. Not activity. Not reasoning. Not calculating. Not
busy behaviour of any kind. Not reading. Not talking. Not making an
effort. Not thinking. Simply *bearing in mind* what it is one needs to know.
And yet those with the courage to tread this path to real discovery are not
only offered practically no guidance on how to do so, they are actively
discouraged and have to set about it in secret, pretending meanwhile to be
diligently engaged in the frantic diversions and to conform with the deadening personal opinions which are continually being thrust upon them.
The Laws of Form 1969 (London: Allen & Unwin)

[Sir] Thomas Browne 1605–1682

3   . . . indeed what reason may not go to Schools to the wisdoms of Bees,
Ants and Spiders? what wise hand teacheth them to doe what reason
cannot teach us? ruder heads stand amazed at those prodigious pieces of
nature, Whales, Elephants, Dromidaries and Camels; these I confesse, are
the Colossus and Majestick pieces of her hand; but in these narrow Engines
there is more Mathematicks, and the civilitie of these little Citizens more
neatly sets forth the wisedome of their Maker.
[Browne's writings are full of curious pre- and proto-scientific learning]
Religio Medici I, 15

4   God is like a skilful Geometrician.
[*cf* Plutarch *Symposiaes* viii, 2: How Plato is to be understood when he saith: That God
continually is exercised in Geometry. It is not, however, in Plato's works]
Religio Medici I, 16

5   All things began in Order, so shall they end, and so shall they begin again,
according to the Ordainer of Order, and the mystical mathematicks of the
City of Heaven.
Hydrotaphia and The Garden of Cyrus 1896 (London: Macmillan)

6   Sure there is music even in the beauty, and the silent note which Cupid
strikes, far sweeter than the sound of an instrument. For there is music
wherever there is harmony, order and proportion; and thus far we may
maintain the music of the spheres; for those well ordered motions, and
regular paces, though they give no sound unto the ear, yet to the understanding they strike a note most full of harmony.
Religio Medici II, 9

1   Thus is Man that great and true Amphibian whose nature is disposed to
    live . . . in divided and distinguished worlds.
    Religio Medici I, 34

2   What song the Syrens sang, or what name Achilles assumed when he hid
    himself among women, though puzzling questions, are not beyond all
    conjecture. [Asked first by Tiberius. Suetonius *Tiberius* LXX]
    [For proposed answers see Robert Graves *The White Goddess* 1948]
    Urn Burial 1658, ch 5

Giordano Bruno 1548–1600

3   *Se no è vero ma è ben trovato.*
    It may not be true but it is well contrived.
    Attributed

John Buchan [Lord Tweedsmuir] 1875–1940

4   To live for a time close to great minds is the best kind of education.
    Memory Hold the Door 1940

Robert Buchanan 1841–1901

5   Alone at nights, I read my Bible more and Euclid less.
    An Old Dominie's Story

Buddha *ca* 563–483 BC

6   All composite things decay. Strive diligently.
    [His last words]

Ludwig Buechner 1824–1899

7   *Ohne Phosphor, kein Gedanke.*
    Without phosphorus there would be no thoughts.
    Attributed

Georges Leclerc [Comte de] Buffon 1707–1788

8   . . . all the work of the crystallographers serves only to demonstrate that
    there is only variety everywhere where they suppose uniformity . . . that in
    nature there is nothing absolute, nothing perfectly regular.
    Histoire Naturelle des Minéraux Paris, 1783–1788, III

9   One can descend by imperceptible degree from the most perfect creature
    to the most shapeless matter, from the best-organised animal to the rough-
    est mineral.
    De la Manière d'étudier et de Traiter l'Histoire Naturelle in *Oeuvres Complètes* Paris, 1774–1791, I

Edward Bulmer-Lytton [Baron Lytton] 1803–1873

10 In science, read, by preference, the newest works; in literature, the oldest.
    Caxtoniana Essay X

Edmund Burke 1729–1797

11 The age of chivalry is gone. That of sophisters, economists and calculators

has succeeded: and the glory of Europe is extinguished for ever.
Reflections on the Revolution in France 1970 (London: Dent)

1   In the groves of *their* academy, at the end of every walk, you see nothing
but the gallows.
Reflections on the Revolution in France 1970 (London: Dent)

[Sir] Frank Macfarlane Burnet 1899–

2   There is virtually nothing that has come from molecular biology that can
be of any value to human living in the conventional sense of what is good,
and quite tremendous possibilities of evil, again in the conventional sense.
The Lancet 1966 **1** 37

Daniel Hudson Burnham 1846–1912

3   Make no little plans, they have no power to stir men's souls.
[Deviser of Lake front park in Chicago]
in Charles Moore *Daniel H Burnham* 1921 (Boston, Mass: Houghton Mifflin)

Robert Burns 1759–1796

4   Facts are chiels that winna ding, an' downa be disputed. [Facts are entities
which cannot be manipulated or disputed.]
A Dream 30

5   I waive the quantum o' the sin
The hazard of concealing
But oh: it hardens a' within
And petrifies the feeling.
Epitaph to Young Friend 6

6   Some books are lies frae end to end,
An' some great lies were never penn'd:
Ev'n ministers they ha'e been kenn'd,
In holy rapture,
A rousing whid at times to vend,
An' nail't wi' Scripture.
[whid = lie]
Death and Doctor Hornbook 1785

Vannevar Bush 1890–

7   The greatest event in the world today is not the awakening of Asia, nor the
rise of communism—vast and portentous as those events are. It is the
advent of a new way of living, due to science, a change in the conditions
of work and the structure of society which began not so very long ago in
the West, and is now reaching out over all mankind.

Samuel Butler 1612–1680

8   In mathematics he was greater
Than Tycho Brahe, or Erra Pater:

For he, by geometric scale,
Could take the size of pots of ale;
Resolve, by sines and tangents straight,
If bread or butter wanted weight;
And wisely tell what hour o' the day
The clock does strike, by Algebra.
Hudibras part 1, 1663

1   A learned society of late,
The glory of a foreign state,
Agreed, upon a summer's night,
To search the Moon by her own light.
The Elephant in the Moon ca 1676

Samuel Butler 1835–1902

2   A hen is only an egg's way of making another egg.
Life and Habit VIII

3   We shall never get people whose time is money to take much interest in atoms.
Notebooks

Herbert Butterfield 1900–

4   It [the Scientific Revolution] outshines everything since the rise of Christianity and reduces the Renaissance and the Reformation to the rank of mere episodes, mere internal displacements, within the system of medieval Christendom . . . . It looms so large as the real origin of the modern world and of the modern mentality that our customary periodisation of European history has become an anachronism and an encumbrance.
The Origins of Modern Science 1949 (London: Bell)

George Gordon [Lord] Byron 1788–1824

5   'Tis a pity learned virgins ever wed
With persons of no sort of education,
Or gentlemen, who, though well born and bred,
Grow tired of scientific conversation.
Don Juan I, XXII

6   'Tis pleasant, sure, to see one's name in print;
A book's a book, although there's nothing in't.
English Bards and Scotch Reviewers line 51

7   When Newton saw an apple fall, he found . . .
A mode of proving that the earth turn'd round
In a most natural whirl, called gravitation,
And thus is the sole mortal who could grapple
Since Adam, with a fall or with an apple.
Don Juan 10, 11

George John Douglas Campbell [8th Duke of Argyll] 1823–1900

1   [*Survival of the fittest*—Herbert Spencer's coinage] Nothing could be
happier than this invention for . . . giving vogue to whatever it might be
supposed to mean . . . . It is the fittest of all phrases to survive.
Organic Evolution Cross-examined 1898

Thomas Campbell 1777–1844

2   O Star-eyed Science! hast thou wandered there,
To waft us home the message of despair?
Pleasures of Hope part 2, line 325

Albert Camus 1913–1960

3   An intellectual is someone whose mind watches itself.
Carnets 1935–1942 1962 (Paris: Gallimard)

Karel Čapek 1860–1927

4   Rossum's Universal Robots.
[The invention of the word 'robot']
R.U.R. 1920 (Oxford: Oxford UP)

Thomas Carlyle 1795–1881

5   Genius . . . means transcendent capacity of taking trouble.
Life of Frederick the Great ch 3

6   In a symbol there is concealment and yet revelation: here therefore, by
Silence and by Speech acting together, comes a double significance.
Sartor Resartus III, iii

7   It is a mathematical fact that the casting of this pebble from my hand alters
the centre of gravity of the universe.
[Does it?]
Sartor Resartus III

8   The Social Science, not a 'gay science' . . . ; no, a dreary, desolate, and
indeed quite abject and distressing one; what we might call, by way of
eminence, the dismal science.
Miscellanies, The Nigger Question

9   Such I hold to be the genuine use of gunpowder; that it makes all men
alike tall.

Alexis Carrel 1873–1944

10   *L'éminence même d'un specialiste le rend plus dangereux.*
The mere eminence of a specialist makes him the more dangerous.
L'homme cet inconnu (Paris: Librairie Plon) ch 1

Lewis Carroll [Charles Lutwidge Dodgson] 1832–1898

11   What I tell you three times is true.
The Hunting of the Snark

1   'Can you do addition?' the White Queen asked. 'What's one and one and
one and one and one and one and one and one and one and one?'
'I don't know,' said Alice, 'I lost count.'
Through the Looking Glass

2   'Why,' said the Dodo, 'the best way to explain it is to do it.'
Alice in Wonderland ch III

3   'Would you tell me, please, which way I ought to go from here?'
'That depends a good deal on where you want to get to,' said the Cat.
'I don't much care where . . . ,' said Alice.
'Then it doesn't matter which way you go,' said the Cat.
'So long as I get somewhere,' Alice added as an explanation.
'Oh, you're sure to do that,' said the Cat, 'If you only walk long enough.'
Alice's Adventures in Wonderland

Charles Frederick Carter 1919–

4   In use of equipment especially, insufficient attention is paid to real costs.
In some cases it would be cheaper not to install the equipment, but when-
ever it is needed to send each student in a separate chauffeur-driven Rolls-
Royce to use the same equipment at another institution.
Universities and Productivity University Conference, Committee of Vice-Chancellors and
Principals and the Association of University Teachers, 21 March 1968

Cato [the Censor] 234–149 BC

5   How could one [Roman] haruspex look another in the face without laugh-
ing?
[A haruspex divined the future from the entrails of animals]
ascribed by Cicero *De Diviniatione* ii, 24

John J Cavanaugh

1   Even casual observation of the daily newspapers and the weekly news magazines, leads a Catholic to ask, where are the Catholic Salks, Oppenheimers, Einsteins?
Time 30 December 1957

Miguel de Cervantes 1547–1616

2   Let us come now to references to authors, which other books contain and yours lacks. The remedy for this is very simple; for you have nothing else to do but look for a book which quotes them all from A to Z, as you say. Then you put this same alphabet into yours. . . . And if it serves no other purpose, at least that long catalogue of authors will be useful to lend authority to your book at the outset.
Don Quixote transl J M Cohen (London: Penguin Classics) Prologue

Paul Cézanne 1839–1906

3   Treat nature in terms of the cylinder, the sphere, the cone, all in perspective.
in Emile Bernard *Paul Cézanne* 1925

George Philip [Air Vice-Marshal] Chamberlain 1905–

4   *Boffin:* A Puffin, a bird with a mournful cry, got crossed with a Baffin, a mercifully obsolete Fleet Air Arm aircraft. Their offspring was a Boffin, a bird of astonishingly queer appearance, bursting with weird and sometimes inopportune ideas, but possessed of staggering inventiveness, analytical powers and persistance. Its ideas, like its eggs, were conical and unbreakable. You push the unwanted ones away, and they just roll back.
in R W Clark *The Rise of the Boffins* 1962 (London: Phoenix House)

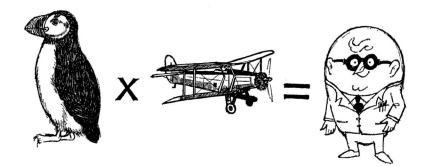

Sébastien Roch Nicholas Chamfort 1740–1794

5   *La Philosophie, ainsi que la Médecine, a beaucoup de drogues, très peu de bons remèdes, et presque point de spécifiques.*
Philosophy, like medicine, has plenty of drugs, few good remedies, and hardly any specific cures.
Maximes et Pensées 1794, 17

Chang Chih-tung 1837–1909

1   Chinese learning as the substance and Western learning for application.
Ch'uan-hsueh p'ien (An exhortation to learning) 1898

Erwin Chargaff 1905–

2   What counts, however, in science is to be not so much the first as the last.
Science 1971 **172** 639

Pierre Charron 1541–1603

3   *La vraye science et le vray étude de l'homme c'est l'homme.*
The true science and study of mankind is man.
De La Sagesse Preface

Geoffrey Chaucer *ca* 1340–1400

4   For out of olde feldes, as men seith,
Cometh al this newe corn fro yeer to yeer;
And out of olde bokes, in good feith,
Cometh all this newe science that men lere.
The Parlement of Foules line 22

5   Little Lewis my son, I have perceived well by certain signs thy ability to
learn sciences touching numbers and proportions; and I also consider thy
earnest prayer specially to learn the Treatise of the Astrolabe . . . . I will
show thee this treatise, divided into five parts, under full easy rules and in
plain English words; for Latin thou knowest as yet but little, my little
son . . . .
Treatise on the Astrolabe Preface

Anton Pavlovich Chekov 1860–1904

6   Under the flag of science, art and persecuted freedom of thought Russia
would one day be ruled by toads and crocodiles the like of which were
unknown even in Spain at the time of the Inquisition.
Letter of 27 August 1888 in Ronald Hingley *Russian Writers and Society* 1967 (London:
Hutchinson)

7   There is no national science just as there is no national multiplication
table; what is national is no longer science.
in V P Pononarev *Mysli o nauke*

Lord Chesterfield [Philip Dormer Stanhope] 1694–1773

8   The pleasure is momentary, the position ridiculous and the expense damn-
able.
[Of sexual intercourse]
Nature 1970 **227** 772

Gilbert Keith Chesterton 1874–1936

9   It isn't that they can't see the solution. It is that they can't see the problem.
The Point of a Pin in *The Scandal of Father Brown* 1935 (London: Cassell)

1   A man must love a thing very much if he not only practices it without any hope of fame and money, but even practices it without any hope of doing it well.

Chinese Proverbs

2   Every truth has its sect
And each sect has its truth.

3   *Yu fang tze shui fang.*
If the bowl be square, the water in it will also be square [indicating the great influence of the prince in moulding the people].

Chou En-lai 1898–1976

4   We must catch up with this advanced level of world science . . . . Only by mastering the most advanced sciences can we insure ourselves of an impregnable national defence, a powerful and up-to-date economy and adequate means to join the Soviet Union and the People's Democracies in defeating the imperial powers, either in peaceful competition or in any aggressive war which the enemy may unleash.
Report on the Question of the Intellectuals 14 January 1956

[Sir] Winston Spencer Churchill 1874–1965

5   . . . I see the absolute truth and explanation of things, but something is left out which upsets the whole, so by a larger sweep of the mind I have to see a greater truth and a more complete explanation which comprises the erring element. Nevertheless, there is still something left out. So we have to take in a still wider sweep. The process continues inexorably. Depth beyond depth of unendurable truth opens.
[Describing his impressions on coming out of an anaesthetic after an accident with a taxi]
My Early Life 1930 (London: Hamlyn)

6   Although personally I am quite content with existing explosives, I feel we must not stand in the path of improvement . . .
[30 August 1941. Minute on report of *MAUD Committee* that it would be possible to make a uranium bomb]
The Second World War 1950 (London: Cassell) vol III

7   Every prophet has to come from civilisation, but every prophet has to go into the wilderness. He must have a strong impression of a complex society and all that it has to give, and then he must serve periods of isolation and meditation. This is the process by which psychic dynamite is made.
[The justification for Sabbatical leave]
Essay on Moses

8   I pass with relief from the tossing sea of Cause and Theory to the firm ground of Result and Fact.
The Story of the Malakand Field Force 1898 (London: Hamlyn)

9   It is a good thing for an uneducated man to read books of quotations.
Roving Commission in *My Early Life* 1930 (London: Hamlyn)

1   Praise up the humanities, my boy. That will make them think that you are broad-minded.
[Advice to R V Jones, his scientific consultant]
Bulletin of the Institute of Physics 1962 **13** 101

2   Scientists should be on tap but not on top.
[Also attributed to Walter Elliot 1888–1958]
in Randolph Churchill *Twenty-one Years* (London: Weidenfeld & Nicolson) Epilogue

3   The Stone Age may return on the gleaming wings of Science.
Science 1969 **163** 1175

4   Unless British science had proved superior to German and unless its strange, sinister resources had been effectively brought to bear in the struggle for survival, we might well have been defeated, and being defeated, destroyed.
The Second World War 1950 (London: Cassell) vol II

Marcus Tullius Cicero 106–43 BC

5   *Omnia quae secundam Naturam fiunt sunt habenda in bonis.*
The works of Nature must all be accounted good.
De Senectute XIX, 71

Arthur Charles Clarke 1917–

6   When a distinguished but elderly scientist states that something is possible, he is almost certainly right. When he states that something is impossible, he is very probably wrong. (Clarke's First Law.)
Profile of the Future 1973 (London: Gollancz)

Rudolf Clausius 1822–1888

7   *Die Energie der Welt ist konstant. Die Entropie der Welt strebt einem Maximum zu.*
The energy of the world is constant. Its entropy tends to a maximum.

William Kingdon Clifford 1845–1879

1   Remember, then, that scientific thought is the guide of action; that the truth at which it arrives is not that which we can ideally contemplate without error, but that which we may act upon without fear; and you cannot fail to see that scientific thought is not an accompaniment of human progress, but human progress itself.
The Common Sense of the Exact Sciences 1885 (Completed by Karl Pearson)

Arthur Hugh Clough 1819–1861

2   And as of old from Sinai's top
God said that God is one,
By Science strict so speaks he now
To tell us there is None.
Earth goes by chemic forces; Heaven's
A Mecanique Celeste.
And heart and mind of human kind
A watch-work as the rest.
The New Sinai

[Sir] Barnett Cocks 1907–

3   A committee is a cul-de-sac down which ideas are lured and then quietly strangled.
New Scientist 8 November 1973

Jean Baptiste Coffinhal 18th Century

4   *La République n'a pas besoin de Savants.*
The Republic has no need of scientists.
[Before ordering the execution of Antoine Lavoisier, May 1794]
Encyclopaedia Britannica 1911, 11th edn, 16-295d

Joel Cohen

5   Physics—envy is the curse of biology.
Science 1971 **172** 675

Simon Cohen 1894–

6   The contributions of Jews to science and invention have been directly in proportion to the amount of freedom they have enjoyed to participate in the contemporary life of the people among whom they lived.
[Director of Research, Brooklyn, New York]
Universal Jewish Encyclopedia 1943, vol 9

Samuel Taylor Coleridge 1772–1834

7   . . . from the time of Kepler to that of Newton, and from Newton to Hartley, not only all things in external nature, but the subtlest mysteries of life and organisation, and even of the intellect and moral being, were conjured within the magic circle of mathematical formulae.
The Theory of Life

8   The first man of science was he who looked into a thing, not to learn

whether it furnished him with food, or shelter, or weapons, or tools, or armaments, or playwiths but who sought to know it for the gratification of knowing.

1   From Avarice thus, from Luxury and War
       Sprang heavenly Science; and from Science Freedom.
       Religious Musings lines 224–5

2   The horned Moon, with one bright star
       Within the nether tip.
       [Scientists have continually reproached poets and painters for their cavalier attitude to the facts of nature]

3   I should not think of devoting less than twenty years to an epic poem. Ten years to collect materials and warm my mind with universal science. I would be a tolerable mathematician. I would thoroughly understand Mechanics; Hydrostatics; Optics and Astronomy; Botany; Metallurgy; Fossilism; Chemistry; Geology; Anatomy; Medicine; then the minds of men, in all Travels, Voyages and Histories. So I would spend ten years; the next five in the composition of the poem, and the next five in the correction of it. So would I write, haply not unhearing of that divine and nightly-whispering voice, which speaks to mighty minds, of predestined garlands, starry and unwithering.
       Letter to Cottle 1796

4   [Sir Thomas Browne discovers] quincunxes in heaven above, quincunxes in earth below, quincunxes in tones, in optic nerves, in roots of trees, in leaves, in everything.
       [What Browne actually described was hexagonal close-packing in a plane. See Browne's *Garden of Cyrus* for a curious neo-Pythagorean speculation on order and structure]
       Encyclopaedia Britannica 1911, 11th edn, 667

1   Men, I think, have to be weighed, not counted.
[On the economics of the Scottish Clearances]

2   Readers may be divided into four classes: 1. Sponges, who absorb all they read, and return it nearly in the same state, only a little dirtied. 2. Sand-glasses, who retain nothing, and are content to get through a book for the sake of getting through the time. 3. Strain-bags, who retain merely the dregs of what they read, and return it nearly in the same state, only a little dirtied. 4. Mogul diamonds, equally rare and valuable, who profit by what they read, and enable others to profit by it also.
Lectures 1811–1812

Alexander Comfort 1920–

3   For poets that have had my luck
Seldom write while they can kiss.
Haste to the wedding 1962 (London: Eyre & Spottiswode)

Arthur Holly Compton 1892–1962

4   The Italian Navigator has reached the New World. And how did he find the Natives? Very friendly.
[Reporting in code by telephone to Conant that the first chain reaction had been initiated]
in Laura Fermi *Atoms in the Family* (Chicago, Ill: University of Chicago Press)

Karl Taylor Compton 1887–1954

5   When I was directing the research work of students in my days at Princeton University, I always used to tell them that if the result of a thesis problem could be forseen at its beginning it was not worth working at.
Hearings on Science Legislation USA, 1945, p623 in D S Greenberg *The Politics of Pure Science* 1967 (New York: New American Library) © D S Greenberg, 1967

Auguste Comte 1798–1857

6   In mathematics we find the primitive source of rationality; and to mathematics must the biologists resort for means to carry out their researches.
Positive Philosophy

7   Men are not allowed to think freely about chemistry and biology, why should they be allowed to think freely about political philosophy?

8   *Prévoir pour pouvoir.*
Foreknowledge is power.

9   To understand a science it is necessary to know its history.
Positive Philosophy

[Marquis de] Condorcet 1743–1794

10   All errors in government and in society are based on philosophic errors

which, in turn, are derived from errors in natural science.
Report and Project of a Decree on the General Organisation of Public Instruction

1   There should exist for all societies a science of maintaining and extending
their happiness; this is what has been called *l'art social*. This science, to
which all others are contributors, has not been treated as a whole. The
science of agriculture, the science of economics, the science of government
. . . are only portions of this greater science. These separate sciences will
not reach their complete development until they have been made into a
well-organised whole . . . . And this result will be obtained sooner if all the
workers are led to follow a constant and uniform method of work.
in *The Validation of Scientific Theories* ed P G Frank, 1961 (New York: Macmillan)

Confucius 551–479 BC

2   The Master said: 'I would not have him to act with me, who will unarmed
attack a tiger, or cross a river without a boat, dying without any regret.
My associate must be the man who proceeds to action full of solicitude,
who is fond of adjusting his plans, and then carries them into execution.'
The Analects in *Sacred Books of the East* transl J Legge, *ca* 1895 (Oxford: Oxford UP)

3   Study the past if you would divine the future.
The Analects in *Sacred Books of the East* transl J Legge, *ca* 1895 (Oxford: Oxford UP)

4   To no one but the Son of Heaven does it belong to order ceremonies, to
fix the measures, and to determine the written characters. Now, over the
kingdom, carriages have all wheels of the same size; all writing is with the
same characters; and for all conduct there are the same rules.
Doctrine of the Mean in *Sacred Books of the East* transl J Legge, *ca* 1895 (Oxford: Oxford UP)

John Constable 1776–1837

5   Painting is a science and should be pursued as an inquiry into the laws of
nature.

Stuart A Copans

6   Why, dear colleagues, must our findings
Now be put in sterile bindings?
Once physicians wrote for recreation.
Our great teachers through the ages,
Fracastoro, and [the] other sages,
Found writing could be fun, like fornication . . .
Perspectives in Biology and Medicine 1973, Winter, p232

Nicolaus Copernicus 1473–1543

7   *Mathemata mathematicis scribuntur.*
Mathematics is written for mathematicians.
De Revolutionibus Preface dedicating the book to Pope Paul III

Le Corbusier 1887–1965

8   *Une maison est une machine-à-habiter.*

A house is a machine for living in.
Vers une architecture 1925, Paris. Transl F Etchells, 1946 (London: Architectural Press)

William Cowper 1731–1800

1   . . . some drill and bore
The solid earth, and from the strata there
Extract a register, by which we learn
That he who made it, and reveal'd its date
To Moses, was mistaken in its age.
Task Book iii *The Garden* Aldine edn, ed J Bruce, 1895 (London: Bell)

Stephen Crane 1871–1900

2   A man said to the universe:
'Sir, I exist:'
'However,' replied the universe,
'The fact has not created in me
A sense of obligation.'
War is kind and other lines 1899 (New York: Knopf)

Francis Albert Eley Crew 1886–1973

3   A few of the results of my activities as a scientist have become embedded
in the very texture of the science I tried to serve—this is the immortality
that every scientist hopes for. I have enjoyed the privilege, as a university
teacher, of being in a position to influence the thought of many hundreds
of young people and in them and in their lives I shall continue to live
vicariously for a while. All the things I care for will continue for they will
be served by those who come after me. I find great pleasure in the thought
that those who stand on my shoulders will see much farther than I did in
my time. What more could any man want?
The Meaning of Death in *The Humanist Outlook* ed A J Ayer, 1968 (London: Pemberton)

Edward Estlin Cummings [e e cummings] 1894–1962

4   O sweet spontaneous
earth how often
has the naughty thumb
of science prodded
thy
beauty
thou answereth them only with
spring.
Tulips and Chimneys 1924 (New York: Seltzer)

Marie [Skłodowska] Curie 1867–1934

5   There are sadistic scientists who hasten to hunt down error instead of establishing the truth.
[A L Mackay's translation]

6   It would be impossible, it would be against the scientific spirit . . . .

Physicists always publish their results completely. If our discovery has a commercial future that is an accident from which we must not profit. And if radium is to be used in the treatment of disease, it seems to me impossible for us to take advantage of that.

[On the patenting of radium. Discussion with her husband, Pierre]
Eve Curie *The discovery of radium* in *Marie Curie* transl V Sheean, 1938 (London: Heinemann)

[Lord] George Nathaniel Curzon 1859–1925

1   The East is a university in which the scholar never takes a degree.

[28 October 1898]
Indian Speeches I, vii. In Kenneth Rose *Superior Person* 1969 (London: Weidenfeld & Nicolson)

2   In the case of Japan I must confess to having departed widely from the accepted model of treatment. There will be found nothing in those pages of the Japan of temples, tea-houses, and bric-a-brac—that infinitesimal segment of the national existence which the traveller is so prone to mistake for the whole, and by doing which he fills the educated Japanese with such unspeakable indignation. I have been more interested in the efforts of a nation, still in pupilage, to assume the manners of the full-grown man, in the constitutional struggles through which Japan is passing, in her relations with foreign Powers, and in the future that awaits her immense ambitions.

Problems of the Far East 1894, Preface to the first edition

The Sixth Dalai Lama 1682–1705

3   This girl was perhaps not born of a mother,
But blossomed in a peach tree:
Her love fades
Quicker than peach-flowers.
Although I know her soft body
I cannot sound out her heart;
Yet we have but to make a few lines on a chart
And the distance of the farthest stars
In the sky can be measured.

Tibet transl G Tucci, 2nd edn, 1973 (London: Elek)

Jean Le Rond D'Alembert 1717–1783

4   *Allez en avant, et la foi vous viendra.*
Push on, and faith will catch up with you.

Salvador Dali 1904–

5   And now the announcement of Watson and Crick about DNA. This is for me the real proof of the existence of God.

in J F C Crick *Of Molecules and Men* 1966 (Washington, DC: University of Washington Press)

Cyril Dean Darlington 1903–

6   It is against the background of conflict and confusion in the relations of science and society that we find ourselves confronted with a crisis in the history of mankind, and particularly in the history of human government.

It is a crisis arising from the rapidly increasing power given to man by science. It is a crisis such as we are accustomed to leave to the arbitrement of sectional interests supported by shouts and cries. But it is one to which scientific inquiry can provide a solution. For the fundamental problem of government is one that can be treated by exact biological methods. It is the problem of the character and the causation of the differences that exist among men, among the races, classes, and individuals which compose mankind. The little passions and prejudices we have been discussing so far fade into nothingness in face of the gigantic errors and illusions that can be, and are being, mobilized to defeat or pervert scientific truth in this field.
Moncure Conway Memorial Lecture 1948 (London: Watts)

1   Mankind . . . will not willingly admit that its destiny can be revealed by the breeding of flies or the counting of chiasmata.
Royal Society Tercentenary Lecture 1960

2   No society can however reach a high development without some kind of priesthood, which organizes a religion designed to govern its breeding behaviour. If breeding behaviour is left at the sole discretion of a governing class, that class, and with it the whole society, is liable to disintegrate.
The Evolution of Man and Society 1969 (London: Allen & Unwin)

Charles Darwin 1808–1882

3   The preservation of favourable variations and the rejection of injurious variations, I call Natural Selection, or Survival of the Fittest. Variations neither useful nor injurious would not be affected by natural selection and would be left a fluctuating element.
Origin of Species

4   Great is the power of steady misrepresentation—but the history of science shows how, fortunately, this power does not long endure.
Origin of Species

5   I see no good reasons why the views given in this volume should shock the religious feelings of anyone.
Origin of Species

6   My mind seems to have become a kind of machine for grinding general laws out of large collections of facts.
Origin of Species

Erasmus Darwin 1731–1802

7   Shall we conjecture that one and the same kind of living filaments is and has been the cause of all organic life?
Zoonomia I, 511

8   Soon shall thy arm, unconquer'd steam: afar
Drag the slow barge, or drive the rapid car;

Or on wide-waving wings expanded bear
The flying chariot through the field of air.
The Botanic Garden I, i, 289

[Sir] Francis Darwin 1848–1925

1   But in science the credit goes to the man who convinces the world, not to the man to whom the idea first occurs.
Eugenics Review 1914 **6** 1

Sushil Chandra Dasgupta

2   Note on Love Waves in a Homogeneous Crust Laid upon a Heterogeneous Medium.
[Title of paper. These waves were called after A E H Love. See *A Treatise on the Mathematical Theory of Elasticity* 1892 (London: Cambridge UP)]
Indian Journal of Theoretical Physics 1953 **1** 121

[Sir] Humphrey Davy 1778–1829

3   The progression of physical science is much more connected with your prosperity than is usually imagined. You owe to experimental philosophy some of the most important and peculiar of your advantages. It is not by foreign conquests chiefly that you are become great, but by a conquest of nature in your own country.
Lecture at the Royal Institution 1809

Stevan Dedijer 1911–

4   The fruitful pursuit of scientific truth and its application, once discovered, is not just a matter of talented individuals well trained in foreign universities and supplied with the equipment they desire. These are very import-

ant, but the cultivation of science is a collective undertaking [written as 'understanding'] and success in it depends on an appropriate social structure. This social structure is the scientific community and its specialised institutions.
Minerva 1963 **2** 81

John Dee 1527–1608

1   A marveilous newtrality have these things mathematicall, and also a strange participation between things supernaturall, immortall, intellectuall, simple and indivisible, and things naturall, mortall, sensible, componded and divisible.
Preface to his edition of *Euclid* 1570

Daniel Defoe *ca* 1660–1731

2   Necessity . . . has so violently agitated the wits of men at this time, that it seems not at all improper . . . to call it, the Projecting Age . . . . The Art of War, which I take to be the highest Perfection of Human Knowledge, is a sufficient proof of what I say, especially in conducting Armies, and in offensive Engines; witness the new ways of Mines, Fougades, Entrenchments, Attacks, Lodgments, and a long *Et Cetera* of New Inventions . . . . But if I would search for a Cause, from whence it comes to pass that this Age swarms with such a multitude of Projectors more than usual; who besides the Innumerable Conceptions which dye in the bringing forth . . . do really every day produce new Contrivances, Engines, and Projects to get Money, never before thought of; if, I say, I would examine whence this comes to pass, it must be thus: The Losses and Depredations which this War brought with it at first were exceeding many . . . [Merchants], prompted by Necessity, rack their Wits for New Contrivances, New Inventions, New Trades, Stocks, Projects, and anything to retrieve the desperate Credit of their Fortunes.
An Essay upon Projects 1697, Introduction

Democritos [of Abdera] *ca* 460–*ca* 370 BC

3   Everything existing in the Universe is the fruit of chance and necessity.
[Taken by Jacques Monod as the title of his book]

4   Nothing exists except atoms and empty space; everything else is opinion.

René Descartes 1596–1650

5   I thought the following four [rules] would be enough, provided that I made a firm and constant resolution not to fail even once in the observance of them.
    The first was never to accept anything as true if I had not evident knowledge of its being so; that is, carefully to avoid precipitancy and prejudice, and to embrace in my judgment only what presented itself to my mind so clearly and distinctly that I had no occasion to doubt it. The second, to divide each problem I examined into as many parts as was feasible, and as was requisite for its better solution. The third, to direct my thoughts in

an orderly way; beginning with the simplest objects, those most apt to be known, and ascending little by little, in steps as it were, to the knowledge of the most complex; and establishing an order in thought even when the objects had no natural priority one to another. And the last, to make throughout such complete enumerations and such general surveys that I might be sure of leaving nothing out.

These long chains of perfectly simple and easy reasonings by means of which geometers are accustomed to carry out their most difficult demonstrations had led me to fancy that everything that can fall under human knowledge forms a similar sequence; and that so long as we avoid accepting as true what is not so, and always preserve the right order of deduction of one thing from another, there can be nothing too remote to be reached in the end, or too well hidden to be discovered.

Discours de la méthode pour bien conduire sa raison et chercher la vérité dans les sciences 1637

1   If we possessed a thorough knowledge of all the parts of the seed of any animal (e.g. man), we could from that alone, by reasons entirely mathematical and certain, deduce the whole conformation and figure of each of its members, and, conversely if we knew several peculiarities of this conformation, we would from those deduce the nature of its seed.
Oeuvres iv, 494

2   It is well to know something of the manners of various peoples, in order more sanely to judge our own, and that we do not think that everything against our modes is ridiculous, and against reason, as those who have seen nothing are accustomed to think.
Discourse on Method part 1

3   A vacuum is repugnant to reason.
Principles of Philosophy 2

Charles Dickens 1812–1870

4   'Yes I have a pair of eyes', replied Sam, 'and that's just it. If they was a pair o' patent double million magnifyin' gas microscopes of hextra power, p'raps I might be able to see through a flight o' stairs and a deal door; but bein' only eyes, you see, my vision's limited'.
The Pickwick Papers

5   When he has learnt that bottinney means a knowledge of plants, he goes and knows 'em. That's our system, Nickleby; what do you think of it?
Nicholas Nickleby

Emily Dickinson 1830–1886

6   Faith is a fine invention
For gentlemen who see;
But microscopes are prudent
In an emergency.
Poems, Second Series ca 1880, XXX

Patric Dickinson 1914–

7   Who were they, what lonely men,

Imposed on the fact of night
The fiction of constellations
And made commensurable
The distances between
Themselves, their loves, and their doubt
Of governments and nations?
Who made the dark stable
When the light was not? Now
We receive the blind codes
Of spaces beyond the span
Of our myths, and a long dead star
May only echo how
There are no loves nor gods
Men can invent to explain
How lonely all men are.
Jodrell Bank in *The World I See* 1960 (London: Chatto & Windus)

Stella Didacus [Diego de Estella] 1524–1578

1   *Pygmaeos gigantum humeris impositos, plusquam ipsos gigantes videre.*
Dwarfs on the shoulders of giants see further than the giants themselves.
Eximii verbi divini CONCIONATORIS ORDINNIS MINORUM Regularis Observantiae Antwerp, 1622. See R K Merton *On the Shoulders of Giants* 1965 (New York: Free Press)

Denis Diderot 1713–1784

2   ... the following definition of an animal: a system of different organic molecules that have combined with one another, under the impulsion similar to an obtuse and muffled sense of touch given to them by the creator of matter as a whole, until each one of them has found the most suitable position for its shape and comfort.
Pensées sur l'interprétation de la nature 1753, LI

3   Do you see this egg? With it you can overthrow all the schools of theology, all the churches of the earth.
Conversations with D'Alembert

4   Men will never be free until the last king is strangled with the entrails of the last priest.
Dithyrambe sur la fête de rois

5   We have three principal means: observation of nature, reflection, and experiment. Observation gathers the facts, reflection combines them, experiment verifies the result of the combination. It is essential that the observation of nature be assiduous, that reflection be profound, and that experimentation be exact. Rarely does one see these abilities in combination. And so, creative geniuses are not common.
Pensées sur l'interprétation de la nature 1753, XV

6   Why should electricity not modify the formation and properties of crystals?
Pensées sur l'interprétation de la nature 1753, XXXIV

1   [Of him] *Je suis bon encyclopédiste,*
Je connais le mal et le bien,
Je suis Diderot à la piste;
Je connais tout et ne crois rien!
I am the true encyclopedist,
I know good and evil,
I am Diderot on the track;
I know everything and believe nothing!
in P Grosclaude *Un Audacieux Message: L'Encyclopédie* Paris, 1951

Diogenes [the Cynic] *ca* 400–*ca* 325 BC

2   For the answer was good that Diogenes made to one that asked him in mockery, How it came to pass that philosophers were the followers of rich men, and not rich men of philosophers. He answered soberly, and yet sharply, Because the one sort knew what they had need of, and the other did not.
in Francis Bacon *Advancement of Learning* I, III, 10

Paul Adrien Maurice Dirac 1902–

3   I think that there is a moral to this story, namely that it is more important to have beauty in one's equations than to have them fit experiment. If Schroedinger had been more confident of his work, he could have published it some months earlier, and he could have published a more accurate equation . . . . It seems that if one is working from the point of view of getting beauty in one's equations, and if one has really a sound insight, one is on a sure line of progress. If there is not complete agreement between the results of one's work and experiment, one should not allow oneself to be too discouraged, because the discrepancy may well be due to minor features that are not properly taken into account and that will get cleared up with further developments of the theory . . . .
Scientific American May 1963

Theodosius Gregorievich Dobzhansky 1900–1975

4   It is possible that there is, after all, something unique about man and the planet he inhabits.
Perspectives in Biology and Medicine 1972, Winter, pp157–75

5   Sir, I do not question your honesty, I question your intelligence.
Nature 11 March 1976

Aelius Donatus 4th Century

6   *Pereant qui ante nos nostra dixerunt.*
To the devil with those who published before us.
[Quoted by St Jerome, his pupil]

John Donne *ca* 1571–1631

7   Let me arrest thy thoughts; wonder with mee,
Why plowing, building, ruling and the rest,
Or most of those arts, whence our lives are blest,

By cursed Cains race invented be,
And blest Seth vext us with Astronomie.
There's nothing simply good, nor ill alone,
Of every quality Comparison
The onely measure is, and judge, Opinion.
The Progress of the Soule 1601, lines 513–20

1   Why grass is green, or why our blood is red
Are mysteries which none have reach'd unto.
Of the Progress of the Soul. The Second Anniversary lines 228–9

2   Then, soul, to thy first pitch work up again;
Know that all lines which circles do contain,
For once that they the centre touch, do touch
Twice the circumference; and be thou such;
Double on heaven, thy thoughts on earth employed;
[*Elements of Euclid* III, 20]
Of the Progress of the Soul. The Second Anniversary lines 435–9

3   And new philosophy call all in doubt;
The element of fire is quite put out;
The sun is lost, and th'earth, and no man's wit
Can well direct him where to look for it.
And freely men confess that this world's spent
When in the planets and the firmament
They seek so many new and see that this
Is crumbled out again to his atomies.
'Tis all in pieces, all coherence gone;
All just supply and all relation.
Prince, subject, father, son are things forgot,
For every man alone thinks he hath got
To be a Phoenix, and that then can be
None of that kind of which he is but he.
Anatomie of the World. First Anniversary 1611, lines 205–18

John Dos Passos 1896–1970

4   All his life Steinmetz was a piece of apparatus belonging to General
Electric.
Proteus in *The 42nd Parallel* (Boston, Mass: Houghton Mifflin)

Mary Douglas 1921–

5   Where there is dirt there is system. Dirt is the byproduct of a systematic
ordering and classification of matter.
Purity and Danger (London: Routledge & Kegan Paul)

[Sir] Arthur Conan Doyle 1859–1930

6   You will, I am sure, agree with me that if page 534 finds us only in the
second chapter, the length of the first one must have been really intolerable.
[The logic is not impeccable, but we must agree with the sentiment]
Sherlock Holmes in *The Valley of Fear* (London: Murray & Cape) ch 1

1   Sherlock Holmes: 'From a drop of
water a logician could infer the
possibility of an Atlantic or a
Niagara without having seen or
heard of one or the other.'
A Study in Scarlet (London: Murray)

2   Sherlock Holmes: 'It is of the
highest importance in the art of
detection to be able to recognise out
of a number of facts which are
incidental and which are vital . . . .
I would call your attention to the
curious incident of the dog in the
night-time. The dog did nothing in the
night-time. That was the curious
incident.'
Silver Blaze in *Memoirs of Sherlock Holmes*

3   When you have eliminated the impossible, whatever remains, however
improbable, must be the truth.
The Sign of Four in *The Adventures of Sherlock Holmes* (London: Murray)

John Dryden 1631–1700

4   [Of George Villiers, second Duke of Buckingham, who 'made the whole
body of vice his study']
A man so various that he seem'd to be
Not one, but all mankind's epitome.
Stiff in opinions, always in the wrong;
Was everything by starts, and nothing long:
But, in the course of one revolving moon,
Was chemist, fiddler, statesman, and buffoon.
Absalom and Achitophel I, line 545

Albrecht Dürer 1471–1528

5   But when great and ingenious artists behold their so inept performances,
not undeservedly do they ridicule the blindness of such men; since sane
judgment abhors nothing so much as a picture perpetrated with no technical knowledge, although with plenty of care and diligence. Now the sole
reason why painters of this sort are not aware of their own error is that
they have not learnt Geometry, without which no one can either be or
become an absolute artist; but the blame for this should be laid upon their
masters, who are themselves ignorant of this art.
The Art of Measurement 1525. Preface to *Of the Just Shaping of Letters* transl R T Nichol,
Book III

Freeman Dyson 1923–

6   Most of the papers which are submitted to the *Physical Review* are rejected,
not because it is impossible to understand them, but because it is possible.

Those which are impossible to understand are usually published.
Innovation in Physics

Maria von Ebner-Eschenbach 1830–1916

1   The manuscript in the drawer either rots or ripens.
Aphorismen

[Meister] Eckhart *ca* 1260–1327

2   The greatest power available to man is not to use it.
Nature 1973 **245** 279

[Sir] Arthur Stanley Eddington 1882–1944

3   I believe there are 15,747,724,136,275,002,577,605,653,961,181,555,468,044, 717,914,527,116,709,366,231,425,076,185,631,031,296 protons in the universe and the same number of electrons.
[$2^{256} \times 136$]
Tarner Lecture 1938. In *The Philosophy of Physical Science* 1939 (London: Cambridge UP)

4   There was once a brainy baboon,
Who always breathed down a bassoon,
For he said, 'It appears
That in billions of years
I shall certainly hit on a tune'.
New Pathways in Science 1935 (London: Cambridge UP) ch 3

5   We used to think that if we knew one, we knew two, because one and one are two. We are finding that we must learn a great deal more about 'and'.

Gösta Carl Henrik Ehrensvard 1905–

6   Consciousness will always be one degree above comprehensibility.
Man on Another World 1965 (Chicago, Ill: University of Chicago Press)

Paul Ralph Ehrlich 1932–

7   The first rule of intelligent tinkering is to save all the parts.
Saturday Review 5 June 1971

Manfred Eigen 1927–

8   A theory has only the alternative of being right or wrong. A model has a third possibility: it may be right, but irrelevant.
The Physicist's Conception of Nature ed Jagdish Mehra, 1973 (Dordrecht: Reidel)

Albert Einstein 1879–1955

9   Common sense is the collection of prejudices acquired by age eighteen.
Scientific American February 1976

10   God does not care about our mathematical difficulties. He integrates empirically.
in L Infeld *Quest* 1942 (London: Gollancz)

1   Dear Sir,   April 23, 1953

Development of Western Science is based on two great achievements; the invention of the formal logical system (in Euclidean geometry) by the Greek philosophers, and the discovery of the possibility to find out causal relationship by systematic experiment (Renaissance). In my opinion one has not to be astonished that the Chinese sages have not made these steps. The astonishing thing is that these discoveries were made at all.

Sincerely yours,

A Einstein

Letter to J E Switzer in D J de S Price *Science since Babylon* 1962 (New Haven, Conn: Yale UP)

2   *Gott würfelt nicht.*
God casts the die, not the dice (Jean Untermeyer).
Albert Einstein: creator and rebel 1973 (London: Hart-Davis, MacGibbon)

3   However, the progress of science presupposes the possibility of unrestricted communication of all results and judgments—freedom of expression and instruction in all realms of intellectual endeavour. By freedom I understand social conditions of such a kind that the expression of opinions and assertions about general and particular matters of knowledge will not involve dangers or serious disadvantages for him who expresses them.
Out of My Later Years 1950 (London: Thames and Hudson)

1   How is it possible to control man's mental evolution so as to make him proof against the psychoses of hate and destructiveness. Here I am thinking by no means only of the so-called uncultured masses. Experience proves that it is rather the so-called Intelligentzia that is most apt to yield to these disastrous collective suggestions, since the intellectual has no direct contact with life in the raw but encounters it in its easiest synthetic form—the printed page.
Letter to Sigmund Freud

2   The whole of science is nothing more than a refinement of everyday thinking.
Out of My Later Years 1950 (London: Thames and Hudson)

3   . . . not only to know how nature is and how her transactions are carried through, but also to reach as far as possible the utopian and seemingly arrogant aim of knowing why nature is thus and not otherwise . . . .
Festschrift für Aurel Stodola 1929 (Zurich: Orell Füssli Verlag)

4   One thing I have learned in a long life: that all our science, measured against reality, is primitive and childlike—and yet it is the most precious thing we have.
Albert Einstein: creator and rebel 1973 (London: Hart-Davis, MacGibbon)

5   The only justification for our concepts is that they serve to represent the complex of our experiences; beyond this they have no legitimacy. I am convinced that the philosophers had a harmful effect upon the progress of scientific thinking in removing certain fundamental concepts from the domain of empiricism, where they are under control, to the intangible heights of the *a priori*—the universe of ideas is just as little independent of the nature of our experiences as clothes are of the form of the human body.
in P A Schlipp *Albert Einstein: Philosopher–Scientist* 1949 (Evanston, Ill: The Library of Living Philosophers)

6   Perfection of means and confusion of goals seem—in my opinion—to characterize our age.
Out of My Later Years 1950 (London: Thames and Hudson)

7   *Raffiniert ist der Herr Gott, aber boshaft ist er nicht.*
God is subtle but he is not bloody-minded.
[Note in the Professor's lounge of the Mathematics Department at Princeton. 'God is slick, but he ain't mean'—Einstein's own translation to Derek Price, 1946]

8   The Temple of Science is a multi-faceted building.
in G Holton *Thematic Origins of Scientific Thought* 1973 (Cambridge, Mass: Harvard UP)

Dwight David Eisenhower 1890–1969

9   In the councils of government we must guard against the acquisition of unwarranted influence, whether sought or unsought, by the military–industrial complex. The potential for the disastrous rise of misplaced power exists and will persist . . . . In holding scientific research and discovery in respect, as we should, we must be alert to the equal and opposite danger

that public policy could itself become the captive of a scientific–techno-logical elite.

Farewell Address as President of the USA, 1961

Walter Elliot 1888–1958

1   Force is not to be used to its uttermost. Nor is thought to be pushed to its logical conclusion.

[As Lord High Commissioner of the Church of Scotland]
Address to the General Assembly of the Church of Scotland. *The Scotsman* 22 May 1957

Ralph Waldo Emerson 1803–1882

2   The English mind turns every abstraction it can receive into a portable utensil, or working institution.

Science Policy 1973 **2** 142

3   A foolish consistency is the hobgoblin of little minds, adored by little statesmen and philosophers and divines.

Self-reliance Essay

4   I hate quotations. Tell me what you know.

Journals May 1849

5   If a man . . . make a better mouse-trap than his neighbour, tho' he build his house in the woods, the world will make a beaten path to his door.

[See *The Oxford Dictionary of Quotations* for the quotation's history]

6   An institution is the lengthened shadow of one man.

Self-reliance Essay

1   Things are in the saddle
And ride mankind.
Ode, inscribed to W H Channing

2   Tobacco, coffee, alcohol, hashish, prussic acid, strychnine, are weak dilutions; the surest is time. This cup which nature puts to our lips, has a wonderful virtue, surpassing that of any other draught. It opens the senses, adds power, fills us with exalted dreams which we call hope, love, ambition, science; especially it creates a craving for larger draughts of itself.

Friedrich Engels 1820–1895

3   . . . science progresses in proportion to the mass of knowledge that is left to it by preceding generations, that is under the most ordinary circumstances in geometrical proportion.
in M M Karpov Osnovyyie zakonomernosti razvitiya estestvoznaniya Rostov State University, 1963

4   Freedom is the recognition of necessity.
[The title chosen by J D Bernal for his book of essays]

5   If you think that I have got hold of something here please keep it to yourself. I do not want some lousy Englishman to steal the idea. And it will take a long time to get it into shape.
Letter to Marx in *Selected Writings* ed W O Henderson, 1967 (London: Penguin)

6   In science, each new point of view calls forth a revolution in nomenclature.
Selected Works of Marx and Engels vol 3 edn, 1973 (London: Lawrence & Wishart)

7   In the book [*A Treatise on Natural Philosophy* Oxford, 1867] by these two Scotsmen [W Thomson and P G Tait] thinking is forbidden, only calculation is permitted. No wonder that at least one of them, Tait, is accounted one of the most pious Christians in pious Scotland.
Dialectics of Nature Moscow, 1954

8   Just as Darwin discovered the law of evolution in organic nature, so Marx discovered the law of evolution in human history; he discovered the simple fact, hitherto concealed by an overgrowth of ideology, that mankind must first of all eat and drink, have shelter and clothing, before it can pursue politics, science, religion, art, etc, and that therefore the production of the immediate material means of subsistence and consequently the degree of economic development attained by a given people or during a given epoch, form the foundation upon which the state institutions, the legal conceptions, the art and even the religious ideas of the people concerned have been evolved, and in the light of which these things must therefore be explained, instead of vice versa as had hitherto been the case. But that is not all. Marx also discovered the special law of motion governing the present-day capitalist mode of production and the bourgeois society that this mode of production has created. The discovery of surplus value suddenly threw light on the problem in trying to solve which all previous investigators, both bourgeois economists and socialist critics, had been groping in the dark. Two such discoveries would be enough for one lifetime. Happy the man to whom it is granted to make even one such dis-

covery. But in every single field which Marx investigated—and he investigated very many fields, none of them superficially—in every field, even
in that of mathematics, he made independent discoveries. Such was the
man of science. But this was not even half the man. Science was for Marx
a historically dynamic, revolutionary force. However great the joy with
which he welcomed a new discovery in some theoretical science whose
practical application perhaps it was as yet quite impossible to envisage, he
experienced quite another kind of joy when the discovery involved immediate revolutionary changes in industry and in the general course of
history.
[Funeral oration]
Selected Works of Marx and Engels vol 3 edn, 1973 (London: Lawrence & Wishart)

1   Life is the mode of existence of proteins, and this mode of existence essentially consists in the constant self-renewal of the chemical constituents of
these substances.
Anti-Dühring 1878

2   While natural science up to the end of the last century was predominantly
a *collecting* science, a science of finished things, in our century it is essentially a *classifying* science, a science of processes, of the origin and development of these things and of the interconnection which binds all these
processes into one great whole.
Ludwig Feuerbach 1886

Dennis Joseph Enright 1920–

3   To shoot a man against the National Library wall!
—The East unsheathes its barbarous finger-nail.
In Europe this was done in railway trucks,
Cellars underground, and such sequestered nooks.
[Written while Visiting Professor in Singapore]
An Unfortunate Poem (II) Warm Protest in *Addictions* 1962 (London: Chatto & Windus)

Euclid [of Alexandria] *ca* 4th Century BC

4   A youth who had begun to read geometry with Euclid, when he had learnt
the first proposition, inquired, 'What do I get by learning these things?'
So Euclid called a slave and said, 'Give him threepence, since he must
make a gain out of what he learns.'
in Stobaeus *Extracts*

Leonard Euler 1707–1783

5   Madam, I have just come from a country where people are hanged if they
talk.
[In Berlin, excusing his taciturnity in conversation with the Queen Mother of Prussia, on his
return from Russia]
in A Vucinich *Science in Russian Culture* 1965 (London: Peter Owen)

Henri Jean Fabre 1823–1915

6   History records the names of royal bastards, but cannot tell us the origin
of wheat.

John King Fairbank 1907–

1   The question now is: Can we understand our stupidity? This is a test of intellect, not of character.
[Professor of Chinese at Harvard]
The Observer 4 May 1975

Michael Faraday 1791–1867

2   [On being offered the Presidency of the Royal Society] Tyndall, I must remain plain Michael Faraday to the last; and let me now tell you, that if I accepted the honour which the Royal Society desires to confer upon me, I would not answer for the integrity of my intellect for a single year.
Tyndall's life in *Experimental Researches in Electricity* (New York: Dover)

3   One day, Sir, you may tax it.
[To Mr Gladstone, the Chancellor of the Exchequer, who asked about the practical worth of electricity]
in R A Gregory *Discovery* 1918 (London: Macmillan)

4   Work, Finish, Publish.
[Benjamin Franklin said much the same]

Haneef A Fatmi and **R W Young**

5   Intelligence is that faculty of mind, by which order is perceived in a situation previously considered disordered.
Nature 1970 **228** 97

Feng-shen Yin-Te 1771–1810

6   With a microscope you see the surface of things. It magnifies them but does not show you reality. It makes things seem higher and wider. But do not suppose you are seeing things in themselves.
The Microscope 1798. See *Report of the Librarian of Congress* 1937

Lawrence Ferlinghetti 1919–

7   Constantly risking absurdity
                              and death
            whenever he performs
                        above the heads
                              of his audience
      the poet like an acrobat
                        climbs on rime
                              to a high wire of his own making . . .
            For he's the super realist
                        who must perforce perceive
                  taut truth
                        before the taking of each stance or step
            in his supposed advance
                        toward that still higher perch
      where Beauty stands and waits
                        with gravity
                              to start her death-defying leap . . .
A Coney Island of the Mind 1958 (New York: New Directions) © Lawrence Ferlinghetti, 1958

Pierre de Fermat 1601–1665

1   [In the margin of his copy of Diophantus' *Arithmetica*, Fermat wrote] To divide a cube into two other cubes, a fourth power or in general any power whatever into two powers of the same denomination above the second is impossible, and I have assuredly found an admirable proof of this, but the margin is too narrow to contain it.

[This was his famous *Last Theorem*]
transl from Latin in *Source Book of Mathematics* 1929 (Maidenhead: McGraw-Hill)

Enrico Fermi 1901–1954

2   Whatever Nature has in store for mankind, unpleasant as it may be, men must accept, for ignorance is never better than knowledge.

in Laura Fermi *Atoms in the family* (Chicago, Ill: University of Chicago Press)

Richard Phillips Feynman 1918–

3   Everything is made of atoms. That is the key hypothesis. The most important hypothesis in all of biology, for example, is that everything that animals do, atoms do. In other words, there is nothing that living things do that cannot be understood from the point of view that they are made of atoms acting according to the laws of physics. This was not known from the beginning it took some experimenting and theorizing to suggest this hypothesis, but now it is accepted, and it is the most useful theory for producing new ideas in the field of biology.

The Feynman Lectures on Physics vol 1, 1963 (London: Addison–Wesley)

4   The whole question of imagination in science is often misunderstood by people in other disciplines. They try to test our imagination in the following way. They say, 'Here is a picture of some people in a situation. What do you imagine will happen next?' When we say, 'I can't imagine,' they may think we have a weak imagination. They overlook the fact that whatever we are allowed to imagine in science must be *consistent with everything else we know*; that the electric fields and the waves we talk about are not just some happy thoughts which we are free to make as we wish, but ideas which must be consistent with all the laws of physics we know. We can't allow ourselves to seriously imagine things which are obviously in contradiction to the known laws of nature. And so our kind of imagination is quite a difficult game. One has to have the imagination to think of something that has never been seen before, never been heard of before. At the same time the thoughts are restricted in a straitjacket, so to speak, limited by the conditions that come from our knowledge of the way nature really is. The problem of creating something which is new, but which is consistent with everything which has been seen before, is one of extreme difficulty.

The Feynman Lectures on Physics vol 2, 1963 (London: Addison–Wesley)

[Sir] Ronald Aylmer Fisher 1890–1962

5   Natural selection is a mechanism for generating an exceedingly high degree of improbability.

6   No aphorism is more frequently repeated . . . than that we must ask Nature

few questions, or ideally, one question at a time. The writer is convinced that this view is wholly mistaken. Nature, he suggests, will best respond to a logically and carefully thought out questionnaire; indeed if we ask her a single question, she will often refuse to answer until some other topic has been discussed.

Perspectives in Biology and Medicine 1973, Winter, p180

Michael Flanders 1922–1975

1   One of the great problems of the world today is undoubtedly this problem of not being able to talk to scientists, because we don't understand science; they can't talk to us because they don't understand anything else, poor dears.

Michael Flanders and Donald Swann *At the Drop of a Hat*

Gustave Flaubert 1821–1880

2   Poetry is as exact a science as geometry.

Bernard Le Bovier [Sieur de] Fontenelle 1657–1757

3   *Toute la philosophie n'est fondée que sur deux choses: sur ce qu'on a l'esprit curieux et les yeux mauvais.*
     Science originates from curiosity and weak eyes.
     Entretiens sur la Pluralité des Mondes, Premier Soir

4   A work of morality, politics, criticism . . . will be more elegant, other things being equal, if it is shaped by the hand of geometry.
     Préface sur l'Utilité des Mathématiques et de la Physique 1729

5   Mathematicians are like lovers . . . . Grant a mathematician the least principle, and he will draw from it a consequence which you must also grant him, and from this consequence another.

6   When the heavens were a little blue arch, stuck with stars, methought the universe was too straight and close: I was almost stifled for want of air: but now it is enlarged in height and breadth, and a thousand vortices taken in. I begin to breathe with more freedom, and I think the universe to be incomparably more magnificent than it was before.
     [Written in 1686]

George Fox 1624–1691

7   I came to know God experimentally.
     [Fox was the founder of the Society of Friends]

[Sir] Theodore Fox 1899–

8   We shall have to learn to refrain from doing things merely because we know how to do them.
     [Editor of *The Lancet*]
     The Lancet 1965 **2** 801

A Frankland 1825–1899

1   I am convinced that the future progress of chemistry as an exact science depends very much upon the alliance with mathematics.
American Journal of Mathematics 1878 **1** 349

Benjamin Franklin 1706–1790

2   He snatched lightning from the heavens and sceptres from kings.
[Epitaph on him by Turgot]

3   To study, to finish, to publish.
[See Michael Faraday]

[Sir] James George Frazer 1854–1941

4   But while science has this much in common with magic that both rest on a faith in order as the underlying principle of all things . . . magic differs widely from that which forms the basis of science.
The Golden Bough abridged edn, 1925 (London: Macmillan)

Frederick the Great [King of Prussia] 1712–1786

5   I have no fault to find with those who teach geometry. That science is the only one which has not produced sects; it is founded on analysis and on synthesis and on the calculus; it does not occupy itself with probable truth; moreover it has the same method in every country.
Oeuvres

Sigmund Freud 1856–1939

6   I am not really a man of science, not an observer, nor an experimenter, and not a thinker. I am by temperament nothing but a conquistador . . . with the curiosity, the boldness and the tenacity that belong to that type of person.
in E Jones *Life and Work of Sigmund Freud* 1953 (London: Hogarth Press) vol 1

7   I have no concern with any economic criticisms of the communistic system; I cannot inquire into whether the abolition of private property is advantageous and expedient. But I am able to recognize that psychologically it is founded on an untenable illusion. By abolishing private property one deprives the human love of aggression of one of its instruments . . . . This instinct did not arise as the result of property; it reigned almost supreme in primitive times when possessions were still extremely scanty.
An Outline of Psychoanalysis (last posthumous essays). In R Ardrey *The Territorial Imperative* 1967 (London: Collins)

8   My life and work has been aimed at one goal only: to infer or guess how the mental apparatus is constructed and what forces interplay and counteract in it.
see E Jones *Life and Work of Sigmund Freud* 1953 (London: Hogarth Press) vol 1

9   [Poets] are masters of us ordinary men, in knowledge of the mind, because they drink at streams which we have not yet made accessible to science.

Robert Frost 1874–1963

1   Did I see it go by,
That Millikan mote?
Well, I said that I did.
I made a good try.
But I'm no one to quote.
If I have a defect
It's a wish to comply
And see as I'm bid.
I rather suspect
All I saw was the lid
Going over my eye,
I honestly think
All I saw was a wink.

[On being asked to look at Millikan's oil-drop experiment for determining the charge on the electron]
A Wish to Comply in *Complete Poems of Robert Frost* 1951 (London: Cape and New York: Holt, Rinehart and Winston) © Robert Frost (© Lesley Frost Ballentine)

2   Some say the world will end in fire,
Some say in ice.
From what I've tasted of desire
I hold with those who favour fire.

Fire and Ice in *The Poetry of Robert Frost* 1969 (London: Cape and New York: Holt, Rinehart and Winston) © Robert Frost (© Lesley Frost Ballentine)

3   The telescope at one end of his beat,
And at the other end the microscope,
Two instruments of nearly equal hope, . . .

The Bear in *The Poetry of Robert Frost* 1969 (London: Cape and New York: Holt, Rinehart and Winston) © Robert Frost (© Lesley Frost Ballentine)

Richard Buckminster Fuller 1895–

4   I am a passenger on the spaceship, Earth.

Operating Manual for Spaceship Earth 1969 (New York: Pocket Books). See also Barbara Ward *Spaceship Earth* 1966 (New York: Columbia UP)

Dennis Gabor 1900–

5   The most important and urgent problems of the technology of today are no longer the satisfactions of the primary needs or of archetypal wishes, but the reparation of the evils and damages wrought by the technology of yesterday.

Innovations: Scientific, Technological and Social 1970 (Oxford: Oxford UP)

6   Short of a compulsory humanistic indoctrination of all scientists and engineers, with a 'Hippocratic oath' of never using their brains to kill people, I believe that the best makeshift solution at present is to give the alpha-minuses alternative outlets for their dangerous brain-power, and this may well be provided by space research.

Inventing the Future (London: Secker & Warburg)

1   Till now man has been up against Nature; from now on he will be up against his own nature.
Inventing the Future (London: Secker & Warburg)

John Kenneth Galbraith 1908–

2   The real accomplishment of modern science and technology consists in taking ordinary men, informing them narrowly and deeply and then, through appropriate organisation, arranging to have their knowledge combined with that of other specialised but equally ordinary men. This dispenses with the need for genius. The resulting performance, though less inspiring, is far more predictable.
The New Industrial State 1967 (London: Hamish Hamilton) © J K Galbraith, 1967

Galileo Galilei 1564–1642

3   I, Galileo Galilei, son of the late Vicenzio Galilei of Florence, aged seventy years, being brought personally to judgment, and kneeling before you, Most Eminent and Most Reverend Lords Cardinals, General Inquisitors of the Universal Christian Commonwealth against heretical depravity, having before my eyes the Holy Gospels which I touch with my own hands, swear that I have always believed, and, with the help of God, will in future believe, every article which the Holy Catholic and Apostolic Church of Rome holds, teaches, and preaches. But because I have been enjoined, by this Holy Office, altogether to abandon the false opinion which maintains that the Sun is the centre and immovable, and forbidden to hold, defend, or teach, the said false doctrine in any manner . . . I am willing to remove from the minds of your Eminences, and of every Catholic Christian, this vehement suspicion rightly entertained towards me, therefore, with a sincere heart and unfeigned faith, I abjure, curse, and detest the said errors and heresies, and generally every other error and sect contrary to the said Holy Church; and I swear that I will never more in future say, or assert anything, verbally or in writing, which may give rise to a similar suspicion of me; but that if I shall know any heretic, or anyone suspected of heresy, I will denounce him to this Holy Office, or to the Inquisitor and Ordinary of the place in which I may be. I swear, moreover, and promise that I will fulfil and observe fully all the penances which have been or shall be laid on me by this Holy Office. But if it shall happen that I violate any of my said promises, oaths, and protestations (which God avert), I subject myself to all the pains and punishments which have been decreed and promulgated by the sacred canons and other general and particular

constitutions against delinquents of this description. So, may God help me, and his Holy Gospels, which I touch with my own hands, I, the above named Galileo Galilei, have abjured, sworn, promised, and bound myself as above; and, in witness thereof, with my own hand have subscribed this present writing of my abjuration, which I have recited word for word.
in J J Fahie *Galileo, His Life and Work* 1903

1   In questions of science the authority of a thousand is not worth the humble reasoning of a single individual.
in *Arago's Eulogy of Laplace* Smithsonian Report, 1874

2   Philosophy is written in that great book which ever lies before our gaze— I mean the universe—but we cannot understand if we do not first learn the language and grasp the symbols in which it is written. The book is written in the mathematical language, and the symbols are triangles, circles and other geometrical figures, without the help of which it is impossible to conceive a single word of it, and without which one wanders in vain through a dark labyrinth.
Opera 4

3   Take note, theologians, that in your desire to make matters of faith out of propositions relating to the fixity of sun and earth you run the risk of eventually having to condemn as heretics those who would declare the earth to stand still and the sun to change position—eventually, I say, at such a time as it might be physically or logically proved that the earth moves and the sun stands still.
Dialogue

[Sir] Francis Galton 1822–1911

4   Whenever you can, count.
in *The World of Mathematics* ed J R Newman, 1956 (New York: Simon & Schuster)

5   [Statistics are] the only tools by which an opening can be cut through the formidable thicket of difficulties that bars the path of those who pursue the Science of Man.
in Karl Pearson *The Life, Letters and Labours of Francis Galton* 1914 (London: Cambridge UP)

Mohandas Karamchand Gandhi 1869–1948

6   [Asked, on his arrival in Europe, what he thought of Western civilisation] 'I think it would be an excellent idea'.
Attributed

Karl Friedrich Gauss 1777–1855

7   . . . *durch planmässiges Tattonieren.*
. . . through systematic feeling about.
[Asked on how he came upon his theorems]

8   God does arithmetic.
Attributed

George III 1738–1820

1   I spend money on war because it is necessary, but to spend it on science, that is pleasant to me.
[To Lalande]
in R A Gregory *Discovery* . . . 1918 (London: Macmillan)

Hans Heinrich Gerth and Charles Wright Mills 1908– and 1916–1962

2   Precisely because of their specialization and knowledge, the scientist and technician are among the most easily used and coordinated of groups in modern society . . . the very rigor of their training typically makes them the easy dupes of men wise in political ways.
Character & Social Structure 1954 (London: Routledge & Kegan Paul)

Al-Ghazali 1058–1111

3   There is no hope in returning to a traditional faith after it has once been abandoned, since the essential condition in the holder of a traditional faith is that he should not know that he is a traditionalist.
in E R Dodds *The Greeks and the Irrational* 1951 (Berkeley, Calif: University of California Press)

Edward Gibbon 1737–1794

4   Twenty-two acknowledged concubines, and a library of sixty-two thousand volumes, attested the variety of his inclinations; and from the productions which he left behind him, it appears that the former as well as the latter were designed for use rather than for ostentation. (By each of his concubines, the younger Gordian left three or four children. His literary productions, though less numerous, were by no means contemptible.)
[Roman Emperor, died 238]
The Decline and Fall of the Roman Empire

5   The winds and waves are always on the side of the ablest of navigators.
The Decline and Fall of the Roman Empire

William Gilbert 1540–1603

6   Look for knowledge not in books but in things themselves.
De Magnete

Allen Ginsberg 1926–

7   The war is language,
language abused
   for Advertisement
language used
like magic for power on the planet
Black Magic language
formulas for reality—
Communism is a 9 letter word
   used by inferior magicians
with the wrong alchemical formula for transforming
   earth into gold
funky warlocks operating on guesswork,

hand-me-down mandrake terminology.

Wichita Vortex Sutra in *The East-side Scene: American Poetry 1960–65* ed A De Loach, 1972 (New York: Doubleday)

Thomas Favill Gladwin 1917–

1   No style of thinking will survive which cannot produce a usable product when survival is at stake.

[On the navigation of the Puluwat Islanders] *East is a Big Bird: Navigation and Logic on Puluwat Atoll* 1970 (Cambridge, Mass: Harvard UP)

Max Gluckman 1911–

2   A science is any discipline in which the fool of this generation can go beyond the point reached by the genius of the last generation.

Politics, Law and Ritual 1965 (New York: Mentor)

Kurt Goedel 1906–

3   It is impossible to demonstrate the non-contradictoriness of a logical mathematic system using only the means offered by the system itself. [Paraphrased]

[See E Nagel and J R Newman *Goedel's Proof*] *Monatshefte für Mathematik und Physik, Leipzig* 1931, pp173–98

Johann Wolfgang von Goethe 1749–1832

4   (*Laboratorium im Sinne des Mittelalters, weitläufige unbehilfliche Apparate zu phantastischen Zwecken.*)
Wagner: *Es wird ein Mensch gemacht.*
. . . nun lässt sich wirklich hoffen,
Dass, wenn wir aus viel hundert Stoffen
Durch Mischung—denn auf Mischung kommt es an—
Den Menschenstoff gemächlich komponieren,
In einen Kolben verlutieren
Und ihn gehörig kohobieren
So ist das Werk im stillen abgetan.
Es wird! Die Masse regt sich klarer!
Die Überzeugung wahrer, wahrer:
Was man an der Natur Geheimnisvolles pries,
Das wagen wir verständig zu probieren,
Und was sie sonst organisieren liess
Das lassen wir kristallisieren.
Mephistopheles: *Wer lange lebt, hat viel erfahren,*
Nichts Neues kann für ihn auf dieser Welt geschehn.
Ich habe schon in meinen Wanderjahren
Kristallisiertes Menschenvolk gesehn.
(Laboratory, after the style of the Middle Ages: extensive, unwieldy apparatus, for fantastical purposes.)
Wagner: A human being in the making . . . .
Look, there's a gleam—
Now hope may be fulfilled,
That hundreds of ingredients, mixed, distilled—
And mixing is the secret—give us power

The stuff of human nature to compound;
If in a limbeck we now seal it round
And cohobate with final care profound,
The finished work may crown this silent hour.
It works. The substance stirs, is turning clearer.
The truth of my conviction presses nearer:
The thing in Nature as high mystery prized,
This has our science probed beyond a doubt;
What Nature by slow process organised,
That have we grasped, and crystallised it out.
Mephistopheles: He who lives long a host of things will know,
The world affords him nothing new to see.
Much have I seen, in wandering to and fro,
Including crystallised humanity.
Faust II, Akt II. Transl Philip Wayne, 1949 (London: Penguin Classics)

1   *'Ins Innre der Natur'—*
   O du Philister—
   Natur hat weder Kern
   Noch Schale.
   'In the inside of Nature'—
   O you Philistines—
   Nature has neither kernel
   Nor shell.
   Allerdings. Dem Physiker 1819/20

2   'Questions of science,' remarked Goethe, 'are very frequently career questions. A single discovery may make a man famous and lay the foundations
   of his fortunes as a citizen . . . . Every newly observed phenomenon is a
   discovery, every discovery is property. Touch a man's property and his
   passions are easily aroused.'
   in J P Eckerman *Conversations with Goethe* 21 December 1823

3   *Auf theoretischem Feld ist weiter nichts mehr zu finden; Aber der praktische*
   Satz gilt doch; Du kannst, denn du sollst.
   In the theoretical field there is no more to be found; but the practical
   dictum is still valid; you can, for you ought.
   Xenien 1797 (jointly with Schiller)

4   *Der kleine Gott . . . . In jeden Quark begräbt er seine Nase.*
   God pushes his nose into all kinds of rubbish.
   [See *Scientific American* July 1968 for the history of the word 'quark']
   Faust (Frankfurt: Insel-Verlag) Prologue

5   Faust*: Geschrieben steht: 'Im Anfang war das Wort!'*
   Hier stock ich schon! Wer hilft mir weiter fort?
   Ich kann das Wort so hoch unmöglich schätzen,
   Ich muss es anders übersetzen,
   Wenn ich vom Geiste recht erleuchtet bin.
   Geschrieben steht: 'Im Anfang war der Sinn.'
   Bedenke wohl die erste Zeile,

Dass deine Feder sich nicht übereile! Ist es der Sinn, der alles wirkt und schafft?
Es sollte stehn: 'Im Anfang war die Kraft!'
Doch, auch indem ich dieses niederschreibe,
Schon warnt mich was, dass ich dabei nicht bleibe.
Mir hilft der Geist! Auf einmal seh ich Rat
Und schreib getrost: 'Im Anfang war die Tat!'
Faust: 'Tis writ: 'In the beginning was the Word!'
I pause, to wonder what is here inferred?
The Word I cannot set supremely high,
A new translation I will try.
I read, if by the spirit I am taught,
This sense: 'In the beginning was the Thought.'
This opening I need to weigh again,
Or sense may suffer from a hasty pen.
Does Thought create, and work, and rule the hour?
'Twere best: 'In the beginning was the Power!'
Yet, while the pen is urged with willing fingers,
A sense of doubt and hesitancy lingers.
The spirit come to guide me in my need,
I write, 'In the beginning was the Deed!'
Faust I. Transl Philip Wayne, 1949 (London: Penguin Classics)

1   *Doch, der den Augenblick ergreift,*
     Das ist der rechte Mann.
     He who seizes the right moment,
     Is the right man.
     Faust I, iii

2   The history of science is science itself: the history of the individual, the
     individual.
     Mineralogy and Geology

3   Nothing is more terrible than to see ignorance in action.
     Maxims and Reflections I

4   Thus I saw that most men only care for science so far as they get a living
     by it, and they worship error when it affords them a subsistence.
     in J P Eckerman *Conversations with Goethe* 15 October 1825

Brian Carey Goodwin 1931–

5   The discovery of appropriate variables for biology is itself an act of
     creation.
     in *Towards a Theoretical Biology* ed C H Waddington, 1969 (Edinburgh: Edinburgh UP)

Robert Ranke Graves 1895–

6   'The sum of all the parts of Such—
     Of each laboratory scene—

Is such.' While Science means this much
And means no more, why, let it mean!
 But were the science-men to find
Some animating principle
Which gave synthetic Such a mind
Vital, though metaphysical—
To Such, such an event, I think
Would cause unscientific pain:
Science, appalled by thought, would shrink
To its component parts again.
Synthetic Such (London: Watts)

1   Myth, then, is a dramatic shorthand record of such matters as invasions,
migrations, dynastic changes, admissions of foreign cults, and social
reforms.
Introduction to the Larousse Encyclopedia of Mythology 1959 (London: Hamlyn)

2   Thought comes often clad in the strangest clothing:
So Kekulé the chemist watched the weird rout
Of eager atom-serpents writhing in and out
And waltzing tail to mouth. In that absurd guise
Appeared benzene and aniline, their drugs and their dyes.
Difficult Questions, Easy Answers 1972 (London: Cassell)

3   To know only one thing well is to have a barbaric mind: civilization implies
the graceful relation of all varieties of experience to a central humane
system of thought. The present age is peculiarly barbaric: introduce, say,
a Hebrew scholar to an ichthyologist or an authority on Danish place
names and the pair of them would have no single topic in common but
the weather or the war (if there happened to be a war in progress, which is
usual in this barbaric age).

Thomas Gray 1716–1771

4   Alas, regardless of their doom,
The little victims play!
No sense have they of ills to come,
Nor care beyond today.
Ode on a Distant Prospect of Eton College

5   Fair science frown'd not on his humble birth,
And melancholy mark'd him for her own.
Elegy written in a Country Churchyard

[Sir] Richard Arman Gregory 1864–1952

6   Science is not to be regarded merely as a storehouse of facts to be used for
material purposes, but as one of the great human endeavours to be ranked
with arts and religion as the guide and expression of man's fearless quest
for truth.

Richard Langton Gregory 1923–

1   On how so little information controls so much behaviour.
    in *Towards a Theoretical Biology* ed C H Waddington, 1969 (Edinburgh: Edinburgh UP)

Murray Christopher Grieve [Hugh McDiarmid] 1892–

2   Perchance the best chance of reproducing the ancient Greek temperament
    would be to cross the Scots with the Chinese.
    Lucky Poet 1943 (London: Methuen)

[General] Leslie Richard Groves 1896–1970

3   [Officer in command of the US Atomic Bomb Installations] Compart-
    mentalization of knowledge, to me, was the very heart of security. My rule
    was simple and not capable of misinterpretation—each man should know
    everything he needed to know to do his job and nothing else.
    [Almost the classic recipe for preventing originality]
    Now it can be told 1962 (New York: Harper & Row)

Ernesto [Che] Guevara 1928–1967

4   When asked whether or not we are Marxists, our position is the same as
    that of a physicist or a biologist who is asked if he is a 'Newtonian', or if
    he is a 'Pasteurian'.
    in *Radical Currents in Contemporary Philosophy* ed David DeGrood, 1971 (St Louis, Mo: Warren
    Green)

Ernst Heinrich Haeckel 1834–1919

5   God . . . [is] . . . a gaseous vertebrate.
    The Riddle of the Universe

6   Ontogeny recapitulates phylogeny. [In full] Ontogenesis, or the develop-
    ment of the individual is a short and quick recapitulation of phylogenesis,
    of the development of the tribe to which it belongs, determined by the
    laws of inheritance and adaptation.
    The History of Creation 1868

[Lord] Hailsham [Quintin Hogg] 1907–

7   None the less I am, so far as I know, the first, and possibly the only
    Minister for Science (or of Science for that matter) in the Universe . . .
    [There was at least one other—in India]
    Science and Politics 1963 (London: Faber & Faber)

John Burdon Sanderson Haldane 1892–1964

8   Cancer's a Funny Thing:
    I wish I had the voice of Homer
    To sing of rectal carcinoma,
    Which kills a lot more chaps, in fact,
    Than were bumped off when Troy was sacked . . . .
    [Written while mortally ill with cancer]
    in Ronald Clark *JBS* 1968 (London: Hodder & Stoughton)

1   The conservative has but little to fear from the man whose reason is the servant of his passions, but let him beware of him in whom reason has become the greatest and most terrible of the passions.
Daedalus, or science and the future 1923 (London: Kegan Paul)

2   I have no doubt that in reality the future will be vastly more surprising than anything I can imagine. Now my own suspicion is that the universe is not only queerer than we suppose, but queerer than we can suppose.
Possible Worlds and Other Papers 1927 (London: Chatto & Windus)

3   I'd lay down my life for two brothers or eight cousins.
New Scientist 8 August 1974

4   In scientific thought we adopt the simplest theory which will explain all the facts under consideration and enable us to predict new facts of the same kind. The catch in this criterion lies in the word 'simplest'. It is really an aesthetic canon such as we find implicit in our criticisms of poetry or painting. The layman finds such a law as $dx/dt = K(d^2x/dy^2)$ much less simple than 'it oozes', of which it is the mathematical statement. The physicist reverses this judgment, and his statement is certainly the more fruitful of the two, so far as prediction is concerned. It is, however, a statement about something very unfamiliar to the plain man, namely, the rate of change of a rate of change.
Science and theology as art forms in *Possible Worlds* 1927 (London: Chatto & Windus)

5   Religion is a way of life and an attitude to the universe. It brings man into closer touch with the inner nature of reality. Statements of fact made in its name are untrue in detail, but often contain some truth at their core. Science is also a way of life and an attitude to the universe. It is concerned with everything but the nature of reality. Statements of fact made in its name are generally right in detail, but can only reveal the form and not the real nature of existence. The wise man regulates his conduct by the theories both of religion and science. But he regards these theories not as statements of ultimate fact, but as art forms.
Science and theology as art forms in *Possible Worlds* 1927 (London: Chatto & Windus)

6   Religion is still parasitic in the interstices of our knowledge which have not yet been filled. Like bed-bugs in the cracks of walls and furniture, miracles lurk in the lacunae of science. The scientist plasters up these cracks in our knowledge; the more militant Rationalist swats the bugs in the open. Both have their proper sphere and they should realise that they are allies.
Science and Life: Essays of a Rationalist 1968 (London: Pemberton and Barrie & Rockliff)

7   A time will however come (as I believe) when physiology will invade and destroy mathematical physics, as the latter has destroyed geometry.
Daedalus, or science and the future 1923 (London: Kegan Paul)

8   We are part of history ourselves, and we cannot avoid the consequences of being unable to think impartially.
Heredity and Politics 1938 (London: Allen & Unwin)

1   Why cannot people learn to speak the truth? I have, I think, taught two, perhaps three, Indian colleagues to do so. It will probably wreck their careers.
in Ronald Clark *JBS* 1968 (London: Hodder & Stoughton)

Stephen Hales 1677–1761

2   Since we are assured that the all-wise Creator has observed the most exact proportions, of number, weight and measure, in the make of all things, the most likely way therefore, to get any insight into the nature of those parts of the creation, which come within our observation, must in all reason be to number, weigh and measure.
Vegetable Staticks Introduction

John Hall 17th Century

3   If that this thing we call the world
By chance on atoms was begot
Which though in ceaseless motion whirled
Yet weary not
How doth it prove
Thou art so fair and I in love.
Epicurean Ode

[Sir] William Rowan Hamilton 1805–1865

4    $i^2 = j^2 = k^2 = ijk = -1.$
[Engraved by him on a stone of Brougham Bridge, over the Royal Canal, Dublin, on 16 October 1863, where the idea of quaternions struck him]
See John Milton *Paradise Lost* v, 181: 'ye Elements . . . that in quaternion run'.

Godfrey Harold Hardy 1877–1947

5   . . . a science is said to be useful if its development tends to accentuate the existing inequalities in the distribution of wealth, or more directly promotes the destruction of human life.
A Mathematician's Apology 1941 (London: Cambridge UP)

6   Beauty is the first test; there is no permanent place in the world for ugly mathematics.
A Mathematician's Apology 1941 (London: Cambridge UP)

7   I have never done anything 'useful'. No discovery of mine has made, or is likely to make, directly or indirectly, for good or ill, the least difference to the amenity of the world . . . . Judged by all practical standards, the value of my mathematical life is nil; and outside mathematics it is trivial anyhow. I have just one chance of escaping a verdict of complete triviality, that I may be judged to have created something is undeniable; the question is about its value.
A Mathematician's Apology 1941 (London: Cambridge UP)

8   There is no scorn more profound, or on the whole more justifiable, than that of the men who make for the men who explain. Exposition, criticism,

appreciation, is work for second-rate minds.
A Mathematician's Apology 1941 (London: Cambridge UP)

[Sir] William Bate Hardy 1864–1934

1   [To Sir Henry Tizard] You know, this applied science is just as interesting as pure science, and what's more it's a damned sight more difficult.
Sir Henry Tizard *Haldane Memorial Lecture* Birkbeck College, University of London, 1955

Herbert Amory Hare 1862–1931

2   At first it is impossible for the novice to cast aside the minor symptoms, which the patient emphasises as his major ones, and to perceive clearly that one or two facts that have been belittled in the narration of the story of the illness are in reality the stalk about which everything in the case must be made to cluster.
Practical Diagnosis 1899 (Philadelphia, Pa: Lea Bros)

William Harvey 1578–1657

3   *Ex ovo omnia.*
Everything from an egg.
De Generatione Animalium London, 1651, Frontispiece. See I B Cohen *Changing Perspectives in the History of Science. Essays in Honour of Joseph Needham* ed M Teich and R Young, 1973 (London: Heinemann)

H G [Blondie] Hasler 1914–

4   You cannot have the success without the failures.
[Organiser of the single-handed Atlantic yacht race]
The Observer 7 July 1968

Stephen William Hawking 1942–

5   God not only plays dice. He also sometimes throws the dice where they cannot be seen.
[See Albert Einstein]
Nature 1975 **257** 362

Friedrich August von Hayek 1899–

6   There are no better terms available to describe this difference between the approach of the natural and the social sciences than to call the former 'objective' and the latter 'subjective'.
The Counter-Revolution of Science 1952 (New York: Free Press)

Oliver Heaviside 1850–1925

7   [Criticised for using formal mathematical manipulations, without understanding how they worked] Should I refuse a good dinner simply because I do not understand the processes of digestion?
[The inventor of the operational calculus and predictor of the Cherenkov effect. Now at last his reputation is increasing to a juster estimate]

Ernst Heinkel 1888–1958

8   [Of Dr Ferdinand Porsche, the automobile engineer] He is a very amiable

man but let me give you this advice. You must shut him up in a cage with seven locks and let him design his engine inside it. Let him hand you the blueprints through the bars. But for heaven's sake don't ever let him see the drawing or the engine again. Otherwise he'll ruin you.

[Modifications to engineering designs are what cost the money]
He 1000 1956 (London: Hutchinson)

Werner Heisenberg 1901–1976

1   Natural science does not simply describe and explain nature; it is part of the interplay between nature and ourselves; it describes nature as exposed to our method of questioning.

Physics and Philosophy 1959 (London: Allen & Unwin)

2   Science clears the fields on which technology can build.

3   An expert is someone who knows some of the worst mistakes that can be made in his subject, and how to avoid them.

Physics and Beyond ed R N Anshen, 1971 (New York: Harper & Row)

Hermann Ludwig Ferdinand von Helmholtz 1821–1894

4   Whoever, in the pursuit of science, seeks after immediate practical utility, may generally rest assured that he will seek in vain.

Academic discourse Heidelberg, 1862

Heraclitus [of Ephesus] *ca* 550–475 BC

5   We must know that war is common to all and strife is justice, and that all things come into being and pass away through strife.

in G S Kirk *Heraclitus, The Cosmic Fragments* 1954 (London: Cambridge UP)

6   All things are an exchange for fire, and fire for all things, even as wares for gold and gold for wares.

in S F Mason *A History of the Sciences* 1953 (London: Routledge & Kegan Paul)

7   If you do not expect the unexpected, you will not find it; for it is hard to be sought out, and difficult.

in Diels *Fragmente der Vorsokratiker* 1st edn, no. 18

Aleksandr Ivanovich Herzen 1812–1870

8   The circulation of accurate and meaningful natural science ideas is of vital concern to our age. These are abundant in science but scarce in society. They should be rendered accessible to all . . . without education in natural science it is impossible to develop a strong intellect . . . . By placing natural science at the beginning of a course of education we would cleanse the child's mind of all prejudices; we would raise him on healthful food until the time when, strong of intellect . . . and ready, he discovers the world of man and history which opens the door for direct participation in the issues of the day.

9   Man and science are two concave mirrors continually reflecting each other.

in *Science and Humanity* 1968 (Moscow: Znanie)

Hermann Hesse 1877–1962

1   You treat world history as a mathematician does mathematics, in which nothing but laws and formulae exist, no reality, no good and evil, no time, no yesterday, no tomorrow, nothing but an eternal, shallow, mathematical present.
The Glass Bead Game 1943 (London: Cape)

David Hilbert 1862–1943

2   One hears a good deal nowadays of the hostility between science and technology. I don't think that is true, gentlemen. I am quite sure that it isn't true, gentlemen. It almost certainly isn't true. It really can't be true.
Sie haben ja gar nichts mit einander zu tun. [They have nothing whatever to do with one another.]
in J R Oppenheimer *Physics in the Contemporary World* (Cambridge, Mass: Harvard UP)

3   Physics is much too hard for physicists.
in Constance Reid *Hilbert* 1970 (London: Allen & Unwin)

4   *Wir müssen wissen. Wir werden wissen.*
We must know. We will know.
[Speech in Königsberg, 1930. Now on his tomb in Göttingen]
in Constance Reid *Hilbert* 1970 (London: Allen & Unwin)

Joel Henry Hildebrand 1881–

5   A child of the new generation
Refused to learn multiplication.
He said 'Don't conclude
That I'm stupid or rude;
I am simply without motivation.'
Perspectives in Biology and Medicine 1970, Winter, p230

[Sir] Cyril Hinshelwood 1897–1967

6   The creative scientist is in fact usually more concerned with the relations of things to one another than with the precise verbal analysis of what these things are. He seeks a representation of the world which continually grows by an extension or transformation of what is there already. Thus what many scientists are really after is the adventure of discovery itself.
British Association for the Advancement of Science Presidential Address, Cambridge, 1965

Hippocrates [of Cos] *ca* 460–*ca* 357 BC

7   Declare the past, diagnose the present, foretell the future.
Epidemics Book I, section 11

8   I swear by Apollo the physician, by Asclepius, by Health, by Panacea and by all the gods and goddesses, making them my witnesses, that I will carry out, according to my ability and judgment, this oath and this indenture. To hold my teacher in this art equal to my own parents; to make him partner in my livelihood; when he is in need of money to share mine with him; to consider his family as my own brothers and to teach them this art, if they

want to learn it, without fee or indenture; to impart precept, oral instruction, and all other instruction to my own sons, the sons of my teacher, and to indentured pupils who have taken the physician's oath, but to nobody else. I will use treatment to help the sick according to my ability and judgment, but never with a view to injury and wrong-doing. Neither will I administer a poison to anybody when asked to do so, nor will I suggest such a course. Similarly, I will not give a woman a pessary to cause abortion. But I will keep pure and holy both in my life and my art. I will not use the knife, not even, verily, on sufferers from stone but I will give place to such as are craftsmen therein. Into whatsoever houses I enter, I will enter to help the sick, and I will abstain from all intentional wrong-doing and harm, especially from abusing the bodies of man or woman, bond or free. And whatsoever I shall see or hear in the course of my profession, as well as outside my profession in my intercourse with men, if it be what should not be published abroad, I will never divulge holding such things to be holy secret. Now if I carry out this oath, and break it not, may I gain for ever reputation among all men for my life and for my art; but if I transgress it and forswear myself, may the opposite befall me.

[Great attention is paid to the trade union aspects of the craft and the demarcation between physicians and surgeons]
The Hippocratic Oath

1   Life is short, the Art long, opportunity fleeting, experience treacherous, judgment difficult. The physician must be ready, not only to do his duty himself, but also to secure the co-operation of the patient, of the attendants and of externals.
Aphorisms I, 1

Thomas Hobbes 1588–1679

2   Nature (the Art whereby God hath made and governs the World) is by the Art of man, as in many other things, so in this imitated, that it can make an Artificial Animal. For seeing life is but a motion of limbs, the beginning whereof is in some principal part within; why may we not say, that all Automata (Engines that move themselves by springs and wheels as doth a watch) have artificial life. For what is the Heart, but a Spring; and the Nerves, but so many Strings; and the joints, but so many Wheels, giving motion to the whole Body, such as was intended by the Artificer? Art goes yet further, imitating that rational and most excellent work of Nature, Man. For by Art is created that great Leviathan called a Commonwealth, or State (in Latin *Civitas*) which is but an artificial man; though of greater stature and strength than the natural, for whose protection and defence it was intended.
Leviathan Introduction, 1651

Eric John Ernest Hobsbawm 1917–

3   There is not much that even the most socially responsible scientists can do as individuals, or even as a group, about the social consequences of their activities.
New York Review of Books XV, 19 November 1970

4   War has been the most convenient pseudo-solution for the problems of

twentieth-century capitalism. It provides the incentives to modernisation and technological revolution which the market and the pursuit of profit do only fitfully and by accident, it makes the unthinkable (such as votes for women and the abolition of unemployment) not merely thinkable but practicable, in the field of policy and administration as well as mass murder. What is equally important, it can re-create communities of men and give a temporary sense to their lives by uniting them against foreigners and outsiders. This is an achievement beyond the power of the private enterprise economy, whose characteristic is that it tends to do precisely the opposite, when left to itself.

The Observer Review 26 May 1968

[Baron] Paul Heinrich Dietrich d'Holbach 1723–1789

1   The unhappiness of man is due to his ignorance of nature.

The System of Nature 1770

Oliver Wendell Holmes 1809–1894

2   A lady's portrait has been known to come out of the finishing-artist's room ten years younger than when it left the camera. But try to mend a stereograph and you will find the difference. Your marks and patches float above the picture and never identify themselves with it. No woman may be declared youthful on the strength of a single photograph; but if the stereoscopic twins say she is young, let her be so acknowledged.

Sun painting in *Atlantic Monthly* July 1861

3   Science is the topography of ignorance.

Medical Essays p211

4   Year after year held the silent toil
     That spread his lustrous coil,
     Still, as the spiral grew,
     He left the past year's dwelling for the new,
     Stole with soft step its shining archway through,
     Built up its idle door,
     Stretched in his last found home, and knew the old no more.

[The Nautilus]

Gerald Holton 1922–

5   During a meeting at which a number of great physicists were to give first-hand accounts of their epoch-making discoveries, the chairman opened the proceedings with the remark: 'Today we are privileged to sit side-by-side with the giants on whose shoulders we stand'.

in D J de S Price *Little Science, Big Science* p1, from Gerald Holton *American Journal of Physics* 1961 **29** 805

Miroslav Holub 1923–

6   With one bold stroke
     he killed the circle, tangent
     and point of intersection in infinity.

  On penalty
of quartering
he banned numbers
from three up.
  Now in Syracuse
he heads a school of philosophers,
Squats on his halberd
for another thousand years
and writes:
one, two
one, two
one, two
one, two.

[Continental armies march 'one, two', rather than 'left, right'.
Holub is a Czechoslovak clinical pathologist whose poems mix scientific, political and
philosophical images]
The Corporal who killed Archimedes in *New Scientist* 24 July 1969

Robert Hooke 1635–1703

1   The business and design of the Royal Society is—to improve the knowledge
of natural things, and all useful Arts, Manufactures, Mechanick practices,
Engynes and Inventions by Experiments—(not meddling with Divinity,
Metaphysics, Moralls, Politicks, Grammar, Rhetorick or Logick) . . . . All
to advance the glory of God, the honour of the King . . . the benefit of his
Kingdom, and the general good of mankind.
1663. In O R Weld *A History of the Royal Society* London, 1848, 1, 146

2   'The True Theory of Elasticity or Springiness' (1676)—*CEIINOSSITTUU.*
[Anagram. Revealed in *De Potentia Restitutiva or of a spring* (1679) as *UT TENSIO, SIC UIS* (as
the extension, so the force). One of the ways of establishing priority in a discovery]

3   The truth is, the science of Nature has been already too long made only a
work of the brain and the fancy. It is now high time that it should return
to the plainness and soundness of observations on material and obvious
things.
Micrographia 1665

Horace [Q Horatius Flaccus] 65–8 BC

4   *Ac ne forte roges, quo me duce, quolare tuter*
Nullius addictus iurare in verba magistri.
    . . . in the word of no master am I bound to believe.
[Hence *nullius in verba*—the motto of the Royal Society]

5   *Persicos odi, puer, apparatus.*
Persian luxury, boy, I hate.
Boy. Those Persians are lousy with apparatus (A L Mackay's translation).
[See *Perspectives in Biology and Medicine* 1972, Summer, pp483–90]
Odes I, 38.1

6   *Naturam expellas furca, tamen usque recurret.*
Though you drive away Nature with a pitchfork she always returns.
Epistles I, x, 24

1   *Omne tulit punctum qui miscuit utile dulci.*
     He gains everyone's approval who mixes the pleasant with the useful.
     Ars Poetica 343

Victor Hugo 1802–1885

2   *Je crois peu à la science des savants bêtes.*
     I don't think much of the science of the beastly scientists.

3   Science says the first word on everything, and the last word on nothing.
     Things of the Infinite: Intellectual Autobiography transl L O'Rourke, 1907 (New York: Funk & Wagnalls)

David Hume 1711–1776

4   If we take in our hand any volume; of divinity or school metaphysics, for
     instance; let us ask, 'Does it contain any abstract reasoning concerning
     quantity or number?' No. 'Does it contain any experimental reasoning
     concerning matter of fact and existence?' No. Commit it then to the flames:
     for it can contain nothing but sophistry and illusion.
     Treatise Concerning Human Understanding

Hu Shih 1891–1962

5   Contact with strange civilizations brings new standards of value, with
     which the native culture is re-examined and re-evaluated, and conscious
     reformation and regeneration are the natural outcome.

6   Even my own name bears witness to the great vogue of evolutionism in
     China. I remember distinctly the morning when I asked my second brother
     to suggest a literary name for me. After only a moment's reflection, he
     said, 'How about the word *shih* (fitness) in the phrase "survival of the
     fittest"?' I agreed, and first using it as a *nom de plume*, finally adopted it in
     1910 as my name.
     Living Philosophies, a Series of Intimate Credos 1931 (New York: Simon & Schuster)

Aldous Leonard Huxley 1894–1963

7   If, O my Lesbia, I should commit,
     Not fornication, dear, but suicide, . . .
     [Elegant verse on a mistaken belief of the ancients that male corpses floated face up and female corpses face down]
     Second Philosopher's Song in *Collected Poetry of Aldous Huxley* 1971 (London: Chatto & Windus)

8   While I have been fumbling over books
     And thinking about God and the Devil and all,
     Other young men have been battling with the days
     And others have been kissing the beautiful women.
     The Life Theoretic in *Collected Poetry of Aldous Huxley* 1971 (London: Chatto & Windus)

9   The difference between a piece of stone and an atom is that an atom is
     highly organised, whereas the stone is not. The atom is a pattern, and the
     molecule is a pattern, and the crystal is a pattern; but the stone, although it
     is made up of these patterns, is just a mere confusion. It's only when life

appears that you begin to get organisation on a larger scale. Life takes the atoms and molecules and crystals; but, instead of making a mess of them like the stone, it combines them into new and more elaborate patterns of its own.
Time Must Have a Stop 1945 (London: Chatto & Windus) ch 14

1   Facts are ventriloquist's dummies. Sitting on a wise man's knee they may be made to utter words of wisdom; elsewhere, they say nothing, or talk nonsense.
Time Must Have a Stop 1945 (London: Chatto & Windus) ch 30

2   'If you look up "Intelligence" in the new volumes of the *Encyclopaedia Britannica*,' he had said, 'you'll find it classified under the following three heads: Intelligence, Human; Intelligence, Animal; Intelligence, Military. My stepfather's a perfect specimen of Intelligence, Military.'
Point Counter Point 1928 (London: Chatto & Windus)

Thomas Henry Huxley 1825–1895

3   The great end of life is not Knowledge but Action.
[Marx was saying much the same thing at the same time]
Technical Education 1877

4   The great tragedy of Science—the slaying of a beautiful hypothesis by an ugly fact.
[There is an interesting philosophical problem as to the nature of the satisfactions obtained from a scientific insight later found to be fallacious]
Biogenesis and Abiogenesis. Collected Essays viii

5   If all the books in the world except the *Philosophical Transactions* were destroyed, it is safe to say that the foundations of physical science would remain unshaken, and that the vast intellectual progress of the last two centuries would be largely, though incompletely, recorded.

6   If only I could break my leg, what a lot of scientific work I could do.
in Cyril Bibby *T H Huxley* (London: Cambridge UP)

7   It is the customary fate of new truths to begin as heresies and to end as superstitions.
The coming of age of the Origin of Species in *Science and Culture* xii

8   It looks as if the scientific, like other revolutions, meant to devour its own children; as if the growth of science tended to overwhelm its votaries; as if the man of science of the future were condemned to diminish into a narrow specialist as time goes on.
in *The Essence of T H Huxley* ed Cyril Bibby, 1967 (London: Macmillan)

9   Science is nothing but trained and organized common sense differing from the latter only as a veteran may differ from a raw recruit: and its methods differ from those of common sense only as far as the guardsman's cut and thrust differ from the manner in which a savage wields his club.
The Method of Zadig in *Collected Essays* IV

1   The State lives in a glasshouse, we see what it tries to do, and all its failures, partial or total, are made the most of. But private enterprise is sheltered under good opaque bricks and mortar. The public rarely knows what it tries to do, and only hears of failures when they are gross and patent to all the world.
Administrative Nihilism 1878

2   That fashioning by Nature of a picture of herself, in the mind of man, which we call the progress of Science . . . .
Nature 1869 **1** 10

3   This seems to be one of the many cases in which the admitted accuracy of mathematical processes is allowed to throw a wholly inadmissible appearance of authority over the results obtained by them. Mathematics may be compared to a mill of exquisite workmanship, which grinds you stuff of any degree of fineness; but, nevertheless, what you get out depends on what you put in; and as the grandest mill in the world will not extract wheat flour from peascods, so pages of formulae will not get a definite result out of loose data.
[No doubt that this elegant statement of the computer scientist's maxim 'garbage in, garbage out', has still earlier versions]
Quarterly Journal of the Geological Society, London 1869 **25** 38

4   Try to learn something about everything and everything about something.
Text on his memorial

5   [Asked by Samuel Wilberforce, Bishop of Oxford, whether he traced his descent from an ape on his mother's or his father's side] If the question is put to me would I rather have a miserable ape for a grandfather or a man highly endowed by nature and possessed of great means of influence and yet who employs those faculties and that influence for the mere purpose of introducing ridicule into a grave scientific discussion—I unhesitatingly affirm my preference for the ape.
[Commemorated by a plate in the University Museum, Oxford]
British Association Meeting University Museum, Oxford, 1860

6   [Of the opening ceremony of Johns Hopkins University] It was bad enough to invite Huxley. It were better to have asked God to be present. It would have been absurd to ask them both.
in C Bibby *Scientist Extraordinary—T H Huxley* 1972 (Oxford: Pergamon)

7   [On first reading Darwin's *Origin of Species*] How extremely stupid not to have thought of that!

Ibn Khaldun 1332–1406

8   Geometry enlightens the intellect and sets one's mind right. All its proofs are very clear and orderly. It is hardly possible for errors to enter into geometrical reasoning, because it is well arranged and orderly. Thus, the mind that constantly applies itself to geometry is not likely to fall into error. In this convenient way, the person who knows geometry acquires

intelligence. It has been assumed that the following statement was written upon Plato's door: 'No one who is not a geometrician may enter our house'.
The Muqaddimah. An Introduction to History N J Dawood's abridgement of F Rosenthal's translation, 1967 (London: Routledge & Kegan Paul)

1   It is a remarkable fact that, with few exceptions, most Muslim scholars both in the religious and in the intellectual sciences have been non-Arabs.
The Muqaddimah. An Introduction to History N J Dawood's abridgement of F Rosenthal's translation, 1967 (London: Routledge & Kegan Paul)

2   Scientific instruction is a craft. This is because skill in a science, knowledge of its diverse aspects, and mastery of it are the result of a habit . . . . The easiest method of acquiring the scientific habit is through acquiring the ability to express oneself clearly in discussing and disputing scientific problems. This is what clarifies their import and makes them understandable.
The Muqaddimah. An Introduction to History N J Dawood's abridgement of F Rosenthal's translation, 1967 (London: Routledge & Kegan Paul)

3   When the Muslims conquered Persia and came upon an indescribably large number of books and scientific papers, Sa'd b. Abi Waqqas wrote to 'Umar b. al-Khattab, asking him for permission to take them and distribute them as booty among the Muslims. On that occasion 'Umar wrote to him: 'Throw them into the water. If what they contain is right guidance, God has given us better guidance. If it is in error, God has protected us against it'. Thus, they [the Muslims] threw them into the water or into the fire, and the sciences of the Persians were lost and did not reach us.
[Abd-ar-Rahman Abu Zayd ibn Muhammad ibn Muhammad ibn Khaldun]
The Muqaddimah. An Introduction to History N J Dawood's abridgement of F Rosenthal's translation, 1967 (London: Routledge & Kegan Paul)

International Business Machines

4   Do not fold, spindle or mutilate.
[A slogan emerging during the student revolution at Berkeley in 1964]
Inscription on IBM card

Isidore of Seville *ca* 7th Century

5   *Tolle numerum omnibus rebus et omnia pereunt.*
Take from all things their number and all shall perish.

Jabir ibn Hayyan [Geber] 8th Century

6   The first essential in chemistry is that thou shouldst perform practical work and conduct experiments, for he who performs not practical work nor makes experiments will never attain to the least degree of mastery. But thou, O my son, do thou experiment so that thou mayest acquire knowledge. Scientists delight not in abundance of material; they rejoice only in the excellence of their experimental methods.
Probably The Discovery of Secrets attributed to Geber, 1892 (London: Geber Society)

Jalal al-Din [Rumi] 1207–1273

7   Man is God's astrolabe. But it requires an astronomer to know the astro-

labe. With that astrolabe what would an ordinary man know of the movements of the circling heavens and the stations of the planets, their influences, transits, and so forth. But in the hands of the astronomer the astrolabe is of great benefit, for he who knows himself knows his Lord. Just as this copper astrolabe is the mirror of the heavens, so the human being is the astrolabe of God. When God causes a man to have knowledge of Him and to know Him and be familiar with Him, through the astrolabe of his own being, he beholds moment by moment and flash by flash the manifestation of God and His infinite beauty and that beauty is never absent from his mirror.

The Discourses of Rumi transl A J Arberry, 1961 (London: Murray)

William James 1842–1910

1   [Of innovations] . . . when a thing was new people said 'It is not true'. Later, when its truth became obvious, people said, 'Anyway, it is not important', and when its importance could not be denied, people said, 'Anyway, it is not new'.

in Lord Ritchie Calder *Leonardo* 1970 (London: Heinemann)

Thomas Jefferson 1743–1826

2   I could more easily believe that two Yankee professors would lie than that stones would fall from heaven.

1807. In *Physics Bulletin* 1968 **19** 225

3   If a due participation of office is a matter of right, how are vacancies to be obtained? Those by death are few: by resignation none.

[Usually quoted as: 'Few die, and none resign']
Letter to a Committee of the Merchants of New Haven, 1801

Francis [Lord] Jeffery 1773–1850

4   'Damn the Solar System. Bad light; planets too distant; pestered with comets; feeble contrivance; could make a better myself.'

in H W Tilman *Mischief in Patagonia* 1966 (London: Cambridge UP)

[Pope] John XXIII 1881–1963

5   Nevertheless, in order to imbue civilisation with sound principles and enliven it with the spirit of the gospel, it is not enough to be illumined with the gift of faith and enkindled with the desire of forwarding a good cause. For this end it is necessary to take an active part in the various organisations and influence them from within. And since our present age is one of outstanding scientific and technical progress and excellence, one will not be able to enter these organisations and work effectively from within unless he is scientifically competent, technically capable and skilled in the practice of his own profession . . .

Encyclical *Pacem in Terris* 10 April 1963, part 5. Official transl by the Vatican Press Office

Samuel Johnson 1709–1784

6   Boswell: 'Is not the Giant's Causeway worth seeing?'

Johnson: 'Worth seeing? Yes; but not worth going to see.'
Boswell's Life of Johnson 12 October 1779

1 Nay, Madam, when you are declaiming, declaim; and when you are calculating, calculate.
Boswell's Life of Johnson 26 April 1776

2 *Network:* Anything reticulated or decussated, at equal distances, with interstices between the intersections.
Dictionary of the English Language

3 The Sciences having long seen their votaries labouring for the benefit of mankind without reward, put up their petition to Jupiter for a more equitable distribution of riches and honour . . . . A synod of the celestials was therefore convened, in which it was resolved, that Patronage should descend to the assistance of the Sciences.
[Science was then beginning to become a profession]
Rambler 29 January 1751, no. 9

4 Sir, I have found you an argument. I am not obliged to find you an understanding.
Boswell's Life of Johnson 19 June 1784

Benjamin Jonson 1573–1637

5 *Surly:* The egg's ordained by Nature to that end, and is a chicken *in potentia.*
Subtle: The same we say of lead and other metals, which would be gold if they had time.
Mammon: And that our art doth further.
The Alchemist II, 2

Bertrand de Jouvenel 1903–

6 Year by year we are becoming better equipped to accomplish the things we are striving for. But what are we actually striving for?
Zukunftspläne. Ritt auf dem Tiger in *Der Spiegel* 1970, no. 1–2

Benjamin Jowett 1817–1893

7 One man is as good as another until he has written a book.
in E A Abbott and L Campbell *Life and Letters of B Jowett* 1897, i

James Joyce 1882–1941

8 I am the greatest engineer who ever lived.
in Marshall McLuhan and Quentin Fiore *War and Peace in the Global Village* (New York: Bantam)

Thomas Hughes Jukes 1906–

9 Very old are the rocks.
The pattern of life is not in their veins.
When the earth cooled, the great rains
came and the seas were filled.

Slowly the molecules enmeshed in
ordered asymmetry.
A billion years passed, aeons of
trial and error.
The life message took form, a spiral,
a helix, repeating itself endlessly,
Swathed in protein, nurtured by
enzymes, sheltered in membranes,
laved by salt water, armored with
lime.
Shells glisten by the ocean marge,
Surf boils, sea mews cry, and the great wind
soughs in the cypress.
[Written by a biologist for his own book]
Molecules and Evolution 1966 (New York: Columbia UP)

Carl Gustav Jung 1875–1961

1   The hypothesis of a collective unconscious belongs to the class of ideas
that people at first find strange but soon come to possess and use as
familiar conceptions. This has been the case with the concept of the
unconscious in general. After the philosophical idea of the unconscious, in
the form presented chiefly by Carus and von Hartman, had gone down
under the overwhelming wave of materialism and empiricism, leaving
hardly a ripple behind, it gradually reappeared in the scientific domain of
medical psychology . . . . A more or less superficial layer of the unconscious
is undoubtedly personal. I call it the personal unconscious. But this personal
unconscious rests upon a deeper layer, which does not derive from personal
experience and is not a personal acquisition but is inborn. This deeper
layer I call the collective unconscious . . . it has contents and modes of
behaviour that are more or less the same everywhere and in all individuals
. . . . The contents of the collective unconscious, on the other hand, are
known as archetypes.
The Archetypes and the Collective Unconscious in *Collected Works* vol 9 (London: Routledge &
Kegan Paul) part 1

2   We can never finally know. I simply believe that some part of the human
Self or Soul is not subject to the laws of space and time.
The Guardian 19 July 1975

Juvenal 1st Century

3   *Grammaticus, rhetor, geometres, pictor, aliptes, augur, schoenobates, medicus,*
magus, omnia novit.
Grammarian, rhetorician, geometer, painter, trainer, soothsayer, rope-
dancer, physician, wizard—he knows everything.
Satires iii, 76

Kalidasa between 200 BC and AD 400

4   If a professor thinks what matters most
Is to have gained an academic post

Where he can earn a livelihood, and then
Neglect research, let controversy rest,
He's but a petty tradesman at the best,
Selling retail the work of other men.

Malavikagnimitra i.17. In *Poems from the Sanskrit* transl John Brough, 1968 (London: Penguin) no. 165. © John Brough, 1968

K'ang Yu-wei 1858–1927

1   In the age of One World, the power of the microscope will be one doesn't know how many times greater than that of [the instrument of] today. [Viewed through the instrument of today] an ant looks like an elephant. [Viewed through the instrument of] the future, the size of a microbe will be like that of the great, skyborne *p'eng* bird.

Ta T'ung Shu: The One-world Philosophy of K'ang Yu-wei 1958 (London: Allen & Unwin)

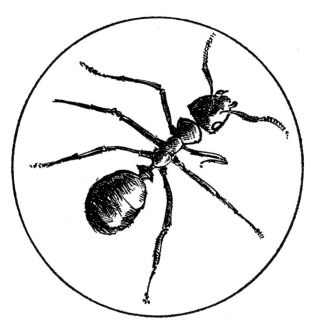

Immanuel Kant 1724–1804

2   Concepts without factual content are empty; sense data without concepts are blind . . . . The understanding cannot see. The senses cannot think. By their union only can knowledge be produced.

3   Two things fill the mind with ever new and increasing admiration and awe, the oftener and more steadily they are reflected on: the starry heavens above me, and the moral law within me.

Critique of Pure Reason

4   *Zweckmässigkeit ohne Zweck.*

Purposiveness without a controlling end.
[A characteristic of biological systems]
Critique of Judgment 1790

Peter Leonidovich Kapitsa 1894–

1   On the flag of contemporary science there should be written in capital letters the word—ORGANISATION.

2   The year that Rutherford died (1938) there disappeared forever the happy days of free scientific work which gave us such delight in our youth. Science has lost her freedom. Science has become a productive force. She has become rich but she has become enslaved and part of her is veiled in secrecy. I do not know whether Rutherford would continue to joke and laugh as he used to.
Science Policy News, London 1969 **1** 33

3   [When asked what the significance of the crocodile carved by Eric Gill on the wall of the Royal Society Mond Laboratory was (crocodile, in fact, was Kapitsa's name for Rutherford)] The crocodile cannot turn its head. Like science, it must always go forward with all-devouring jaws.
in A S Eve *Rutherford* 1939 (London: Cambridge UP)

John Keats 1795–1821

4   Do not all charms fly
At the mere touch of cold philosophy?
There was an awful rainbow once in heaven:
We know her woof, her texture; she is given
In the dull catalogue of common things.
Philosophy will clip an angel's wings,
Conquer all mysteries by rule and line,
Empty the haunted air, and gnomed mine
Unweave a rainbow.
Lamia 1820, II, lines 229–37

Keikitsu 1694–1761

5   *Doko no jishaku mo Yoshiwara e muku.*
From wherever it is the magnet points to the Yoshiwara (gay quarter).
Mutamagawa 1754

Thomas à Kempis 1380–1471

6   The humble knowledge of thyself is a surer way to God than the deepest search after science.
De Imitatione Christi part 1, ch 3

Maurice George Kendall 1907–

7   But I do wish to propound one principle which is, so to speak, a kind of Occam's Electric Razor: We should not invoke any entities or forces to explain mental phenomena if we can achieve an explanation in terms of a

possible electronic computer.
Review of the International Statistical Institute 1966 **34** 1

John Fitzgerald Kennedy 1917–1963

1   First, I believe that this nation should commit itself to achieving the goal, before this decade is out, of landing a man on the moon and returning him safely to earth. No single space project in this period will be more exciting, or more impressive to mankind, or more important for the long-range exploration of space; and none will be so difficult or expensive to accomplish.
[To Congress, 25 May 1961]
in John M Logsdon *The Decision to go to the Moon* 1970 (Cambridge, Mass: The MIT Press)

2   Scientists alone can establish the objectives of their research, but society, in extending support to science, must take account of its own needs.
Address to the National Academy of Sciences, 1963

Hugh Kenner 1923–

3   Each of us carries in his mind a phantom cube, by which to estimate the orthodoxy of whatever we encounter in the world of space.
[This is perhaps more true than Kenner thought. The three semi-circular canals in the ear provide us with built-in Cartesian axes]
Bucky 1973 (New York: Morrow)

Johannes Kepler 1571–1630

4   . . . . Why they are as they are, and not otherwise.
Mysterium Cosmographicum Preface

5   [Title page of *Tertius inveniens*] A warning to sundry Theologos, Medicos, and Philosophos, in particular to D Philippus Feselius, that they should not, in their just repudiation of star-gazing superstition, throw out the child with the bath and thus unknowingly act in contradiction to their profession.
in C G Jung and W Pauli *The Interpretation of Nature and the Psyche* 1955 (London: Routledge & Kegan Paul)

6   It may well wait a century for a reader, as God has waited six thousand years for an observer.
in David Brewster *Martyrs of Science or the Lives of Galileo, Tycho Brahe, and Kepler* 1841

7   *Ubi materia, ibi geometria.*
Where there is matter, there is geometry.

Charles Franklin Kettering 1876–1958

8   [Vice-president of General Motors] Bankers regard research as most dangerous and a thing that makes banking hazardous due to the rapid changes it brings about in industry. (Address, 1927.)
in US National Resources Committee *Technology and Planning* Washington, 1937

John Maynard Keynes 1883–1946

9   [Of him] No one in our age was cleverer than Keynes nor made less

attempt to conceal it.
in R F Harrod *The Life of John Maynard Keynes* 1951 (New York: Harcourt, Brace Jovanovich)

1   The difficulty lies, not in the new ideas, but in escaping the old ones, which ramify, for those brought up as most of us have been, into every corner of our minds.

2   Newton was not the first of the age of reason. He was the last of the magicians, the last of the Babylonians and Sumerians, the last great mind which looked out on the visible and intellectual world with the same eyes as those who began to build our intellectual inheritance rather less than 10 000 years ago.
Address to the Royal Society Club, 1942

3   The avoidance of taxes is the only intellectual pursuit that still carries any reward.

Søren Kiekegaard 1813–1855

4   Knowledge is an attitude, a passion, actually an illicit attitude. For the compulsion to know is just like dipsomania, erotomania, homocidal mania, in producing a character that is out of balance. It is not at all true that the scientist goes after truth. It goes after him.

Rudyard Kipling 1865–1936

5   And your rooms at college was beastly—more like a whore's than a man's.
[A dying shipping tycoon explains his will to his more gently nurtured son]
The Mary Gloster in *The Definitive Edition of Rudyard Kipling's Verse* 1940 (London: Hodder & Stoughton)

6   But remember please, the Law by which we live,
We are not built to comprehend a lie,
We can neither love nor pity nor forgive.
If you make a slip in handling us you die.
The Secret of the Machines in *The Definitive Edition of Rudyard Kipling's Verse* 1940 (London: Hodder & Stoughton)

7   For undemocratic reasons and for motives not of State,
They arrive at their conclusions—largely inarticulate.
Being void of self-expression they confide their views to none;
But sometimes in a smoking room, one learns why things were done.
The Puzzler in *The Definitive Edition of Rudyard Kipling's Verse* 1940 (London: Hodder & Stoughton)

8   Nothing in life has been made by man for man's using
But it was shown long since to man in ages
Lost as the name of the maker of it,
Who received oppression and scorn for his wages—
Hate, avoidance, and scorn in his daily dealings—
Until he perished, wholly confounded.
More to be pitied than he are the wise
Souls which foresaw the evil of loosing

Knowledge or Art before time, and aborted
Noble devices and deep-wrought healings,
Lest offence should arise.
Heaven delivers on earth the Hour that cannot be thwarted,
Neither advanced, at the price of a world or a soul, and its Prophet
Comes through the blood of the vanguards who dreamed—too soon—it
had sounded.
The Eye of Allah in *Debits and Credits* 1926 (London: Methuen)

Aleksander Isaakovich Kitaigorodskii 1914–

1   A first-rate theory predicts; a second-rate theory forbids; and a third-rate
theory explains after the event.
Lecture, IUC Amsterdam, August 1975

Spencer Klaw 1920–

2   It is only at long intervals that the researcher enjoys the feeling (or illusion)
of solid accomplishment that the administrator can enjoy merely by empty-
ing his in-box.
Science 1969 **163** 60

Arthur Koestler 1905–

3   Nobody before the Pythagoreans had thought that mathematical relations
held the secret of the universe. Twenty-five centuries later, Europe is still
blessed and cursed with their heritage. To non-European civilizations, the
idea that numbers are the key to both wisdom and power, seems never to
have occurred.
The Sleepwalkers 1959 (London: Hutchinson)

4   To believe entails no desire to know; everybody reads the Bible but who
reads Flavius Josephus?
The Yogi and the Commissar 1945 (London: Cape)

Roy M Kohn

5   Both horns of a dilemma are usually attached to the same bull: A built-in
impediment to understanding psychoses.
[Title of paper]
Perspectives in Biology and Medicine 1970, Summer, p633

William Lester Kolb 1916–

6   *Science A:* In modern social science usage, the term *Science* denotes the
systematic, objective study of empirical phenomena and the resultant bodies
of knowledge. It is believed by social scientists that their disciplines are
themselves sciences in this sense, and that science as a human activity is
itself an object of social science investigation.
A Dictionary of the Social Sciences ed Julius Gould and William L Kolb, 1964 (Paris: UNESCO/
Tavistock)

Jan Amos Komensky [Comenius] 1592–1670

7   We bid you, then, who are priests in the realm of nature, to press on your

labours with all vigour. See to it that mankind is not for ever mocked by a Philosophy empty, superficial, false, uselessly subtle. Your heritage is a fair Sparta; enrich her with fair equipment, and by making a strict examination both of facts and of opinions concerning them, set an example, as you properly may, to politicians and theologians. He was right who said that a contentious philosophy is the parent of a contentious theology: we must therefore say at once and plainly about Politics that the main political theories on which the present rulers of the world support themselves are treacherous quagmires and the real causes of the generally tottering and indeed collapsing condition of the world. It is for you to show that errors are no more to be tolerated, even though they have the authority of long tradition and are drawn from Adam himself; you must show, not only to theologians, but to the politicians themselves, that everything must be called back to Urim and Thummim, I mean to Light and Truth.

The Way of Light Amsterdam, 1668

The Koran

1   We made from water every living thing.
     Sura 21: 31

2   We have sent down iron, with its mighty strength and diverse uses for mankind, so that Allah may know those who aid Him, though unseen, and help His apostles. Powerful is Allah, and mighty.
     Sura *Iron* 57: 25

Korean Proverb

3   *Kwon un sip nyon i yo, sye nun paik nyon i ra.*
     Power lasts ten years; influence not more than a hundred.

Damodar Dharmonand Kosambi 1907–1966

4   . . . the work remains unique in all Indian literature because of its complete freedom from cant and absence of specious reasoning.
     [On the *Arthashastra* of Kautilya]
     The Culture and Civilisation of Ancient India 1965 (London: Routledge & Kegan Paul)

Tadeusz Kotarbinski 1886–

5   *Inteligent jest to pasozyt wytwarzacy kulture.*
     An intellectual is a parasite that exudes culture.
     in *Yugoslav Dictionary of Quotations*

6   *Tam jest potrebny kontroler, gdzie jest potrezebny kontroler kontrolera.*
     Where there is the need for a controller, a controller of the controller is also needed.
     in *Yugoslav Dictionary of Quotations*

Karl Kraus 1868–1952

7   Science is spectrum analysis. Art is photosynthesis.

Leopold Kronecker 1823–1891

1   *Die ganze Zahl schuf der liebe Gott, alles Übrige ist Menschenwerk.*
     God made the integers, man made the rest.
     Jahresberichte der deutschen Mathematiker Vereinigung book 2, p19. In F Cajori *A History of Mathematics* 1919 (London: Macmillan)

[Prince] Peter Alekseyevich Kropotkin 1842–1921

2   There are not many joys in human life equal to the joy of sudden birth of
     a generalization . . . . He who has once in his life experienced this joy of
     scientific creation will never forget it.

Kuan Yin Tze 8th Century

3   Those who are good at archery learnt from the bow and not from Yi the
     Archer. Those who know how to manage boats learnt from boats and not
     from Wo [the legendary mighty boatman]. Those who can think learnt for
     themselves and not from the Sages.
     in J Needham *Science and Civilization in China* 1956 (London: Cambridge UP)

Otto Vilhelmovich Kuusinen 1881–1964

4   The Marxist–Leninist world outlook stems from science itself and trusts
     science, as long as science is not divorced from reality and practice.
     Fundamentals of Marxism–Leninism Moscow, 1963

Jean de La Bruyère 1645–1696

5   *Les médecins laissent mourir, les charlatans tuent.*
     . The doctors allow one to die, the charlatans kill.
     Les Characteres

Diogenes Laertius 3rd Century BC

6   [When someone asked him his nationality, he replied] cosmopolitan.
     Diogenes Laertius VI, 63

Alphonse Marie Louis Prat de Lamartine 1790–1869

7   It is not I who think, but my ideas which think for me.

Charles Lamb 1775–1834

8   Nothing puzzles me more than time and space; and yet nothing troubles
     me less, as I never think about them.
     Letter to Thomas Manning, 2 January 1806

Frederick William Lanchester 1868–1946

9   The fighting strength of a force may be broadly defined as proportional to
     the square of its numerical strength multiplied by the fighting value of its
     individual units.

     [An engineer of genius and a pioneer in the quantization of social phenomena]
     Aircraft in Warfare 1916 (London: Constable)

Andrew Lang 1844–1912

1   He uses statistics as a drunken man uses lamp-posts—for support rather
     than illumination.

Ray Lankester 1847–1929

2   The fact that we are able to classify organisms at all in accordance with the
     structural characteristics which they present, is due to the fact of their
     being related by descent.
     in D'Arcy Thompson *Growth and Form* 1917 (London: Cambridge UP)

Lao Tze 604–531 BC

3   Clay is moulded to make a vessel, but the utility of the vessel lies in the
     space where there is nothing . . . . Thus, taking advantage of what is, we
     recognise the utility of what is not.
     Tao Te Ching ch 41

4   Nature is not human-hearted (anthropomorphic).
     Tao Te Ching ch 5

5   Of the second-rate rulers, people speak respectfully, saying, 'He has done

this, he has done that'. Of the first-rate rulers they do not say this. They say: 'We have done it all ourselves'.

[There are dozens of translations of Lao Tze. Some, such as this one, stretch the meaning very far, as the text can mean all things to all men. However, the cumulative effect of the text cannot be mistaken and thus perusal of the whole is recommended. There is indeed some evidence that the dialectical method of Hegel was influenced by Chinese traditions. We do not know who supplied this translation, but the sentiment accurately characterises directors of research projects]
Tao Te Ching ch 17

1   A good calculator does not need artificial aids.
Tao Te Ching ch 27

Pierre Simon de Laplace 1749–1827

2   Given for one instant an intelligence which could comprehend all the forces by which nature is animated and the respective positions of the beings which compose it, if moreover, this intelligence were vast enough to submit these data to analysis, it would embrace in the same formula both the movements of the largest bodies in the universe and those of the lightest atom; to it nothing would be uncertain, and the future as the past would be present to its eyes.
Oeuvres vol VII. *Theorié Analytique de Probabilité* 1812–1820, Introduction

3   *Napoleon:* 'You have written this huge book on the system of the world without once mentioning the author of the universe'.
Laplace: 'Sire, I had no need of that hypothesis'.
    Later, when told by Napoleon about the incident, Lagrange commented: 'Ah, but that is a fine hypothesis. It explains so many things'.

Ferdinand Lasalle 1825–1864

4   Whoever obstructs scientific inquiry clamps down the safety valve of public opinion and puts the state in train for an explosion.
Science and the Workingmen 1863

Emanuel Lasker 1868–1941

5   On the chessboard, lies and hypocrisy do not survive long. The creative combination lays bare the presumption of a lie; the merciless fact, culminating in a checkmate, contradicts the hypocrite.
[We find that computer programming imposes the same discipline]

Latin Proverb

6   *Ubi bene ibi patria.*
He makes his home where the living is best.

Johann Caspar Lavater 1741–1801

7   Who in the same given time can produce more than others has *vigour*; who can produce more and better, has *talents*; who can produce what none else can, has *genius*.
Aphorisms on man London, 1788, no. 23

David Herbert Lawrence 1885–1930

1   You can't invent a design. You recognise it, in the fourth dimension. That is, with your blood and your bones, as well as with your eyes.
Phoenix: Art and Morality

Gustave Le Bon 1841–1931

2   At the bidding of a Peter the Hermit millions of men hurled themselves against the East; the words of an hallucinated enthusiast such as Mahomet created a force capable of triumphing over the Graeco–Roman world; an obscure monk like Luther bathed Europe in blood. The voice of a Galileo or a Newton will never have the least echo among the masses. The inventors of genius hasten the march of civilisation. The fanatics and the hallucinated create history.

Henri Le Châtelier 1850–1936

3   Every system in chemical equilibrium, under the influence of a change of any single one of the factors of equilibrium, undergoes a transformation in such direction that, if this transformation took place alone, it would produce a change in the opposite direction of the factor in question. The factors of equilibrium are temperature, pressure and electromotive force, corresponding to the three forms of energy—heat, electricity and mechanical energy.
Recherches sur les Équilibres Chimiques 1888. *Comptes Rendus* 1884 **99** 786

Thomas Andrew Lehrer 1928–

4   In one word he told me the secret of success in mathematics: plagiarize . . . only be sure always to call it please research.
[Song from a very successful gramophone record, *ca* 1960]
Lobachevski

5   And what is it that put America in the forefront of the nuclear nations? And what is it that will make it possible to spend 20 billion dollars of your money to put some clown on the moon? Well, it was good old American know-how, that's what, as provided by good old Americans like Dr Wernher von Braun.
[Gramophone record]
Wernher von Braun on *That Was The Year That Was (TW3 songs and other songs of the year)* 1965

Gottfried Wilhelm Leibnitz 1646–1716

6   The art of discovering the causes of phenomena, or true hypotheses, is like the art of deciphering, in which an ingenious conjecture greatly shortens the road.
New Essays Concerning Human Understanding IV, XII

Philip Lenard 1862–1947

7   No entry to Jews and Members of the German Physical Society.
[Notice on his door]
in K Mendelssohn *The World of Walther Nernst. The Rise and Fall of German Science* 1973 (London: Macmillan)

Vladimir Ilich Lenin 1870–1924

1   Communism is Soviet power plus the electrification of the whole country.
[Slogan promoting the plan of GOELRO—the State Commission for the Electrification of Russia] 1920

2   Given these economic premises it is quite possible, after the overthrow of the capitalists and bureaucrats, to proceed immediately overnight to supersede them in the control of production and distribution, in the work of keeping account of labour and products by the armed workers, by the whole of the armed population. (The question of control and accounting must not be confused with the question of the scientifically trained staff of engineers, agronomists and so on. These gentlemen are working today and obey the capitalists; they will work even better tomorrow and obey the armed workers.)
State and Revolution 1917, ch 5

3   It is absurd to deny the role of fantasy in even the strictest science.
Polnoe Sobranie Sochinenii 5th edn, vol 29

4   Socialism is inconceivable without . . . engineering based on the latest discoveries of modern science.
Left-wing Childishness and the Petty-bourgeois Mentality Moscow, 1968

5   We do not invent, we take ready-made from capitalism; the best factories, experimental stations and academies. We need adopt only the best models furnished by the experience of the most advanced countries.
The Impending Catastrophe 1917

Leonardo da Vinci 1452–1519

6   The function of muscle is to pull and not to push, except in the case of the genitals and the tongue.
in Edward MacCurdy *The Notebooks of Leonardo da Vinci* vol 1. 1938 (London: Cape) ch 3

7   *Il sole no si move.*
The Sun does not move.
Works ed J P and I A Richter (London: Phaidon Press)

8   *Nessuna humana investigazione si pio dimandara vera scienzia s'essa non passa per le matematiche dimonstrazione.*
No human investigation can be called real science if it cannot be demonstrated mathematically.

9   Shun those studies in which the work that results dies with the worker.
in Edward MacCurdy *The Notebooks of Leonardo da Vinci* vol 1, 1938 (London: Cape) ch 1

10   There is no higher or lower knowledge, but one only, flowing out of experimentation.

R L Lesher and G J Howick

11   Eight hundred life spans can bridge more than 50,000 years. But of these

800 people, 650 spent their lives in caves or worse; only the last 70 had any truly effective means of communicating with one another, only the last 6 ever saw a printed word or had any real means of measuring heat or cold, only the last 4 could measure time with any precision; only the last 2 used an electric motor; and the vast majority of the items that make up our material world were developed within the lifespan of the eight-hundredth person.

Assessing Technology Transfer 1966 *NASA Report* SP-5067, pp9–10

Claude Levi-Strauss 1908–

1   *La langue est une raison humaine qui a ses raisons, et que l'homme ne connaît pas.*
Language is human reason, which has its internal logic of which man knows nothing.

La Pensée Sauvage 1962 (Paris: Librairie Plon)

2   No science today can consider the structures with which it has to deal as being more than a haphazard arrangement. That arrangement alone is structured which meets two conditions: that it be a system, ruled by an internal cohesiveness; that this cohesiveness, inaccessible to observation in an isolated system, be revealed in the study of transformations, through which the similar properties in apparently different systems are brought to light.

Leçon inaugurale in *The Scope of Anthropology* transl S O and R A Paul, 1967 (London: Cape)

Cecil Day Lewis 1904–

3   And the mind must sweat a poison
... that, discharged not thence
Gangrenes the vital sense
And makes disorder true.
It is certain we shall attain
No life till we stamp on all
Life the tetragonal
Pure symmetry of the brain.

Transitional Poem in *Collected Poems 1929–1933* 1945 (London: Hogarth Press)

Percy Wyndham Lewis 1884–1957

4   Wherever there is objective truth, there is satire.

Rude Assignment 1950 (London: Hutchinson)

Sinclair Lewis 1885–1951

5   Our American professors like their literature clear, cold, pure and very dead.

Address to the Swedish Academy 1930

William Leybourn 1626–1700

6   But leaving those of the Body, I shall proceed to such Recreations as adorn the Mind; of which those of the Mathematicks are inferior to none.

Pleasure with Profit London, 1694

Sam Lilley

1   The form of society has a very great effect on the rate of inventions and a form of society which in its young days encourages technical progress can, as a result of the very inventions it engenders, eventually come to retard further progress until a new social structure replaces it. The converse is also true. Technical progress affects the structure of society.
Men, Machines and History 1948 (London: Lawrence & Wishart)

Charles Augustus Lindberg 1902–1974

2   The tragedy of scientific man is that he has found no way to guide his own discoveries to a constructive end. He has devised no weapon so terrible that he has not used it. He has guarded none so carefully that his enemies have not eventually obtained it and turned it against him. His security today and tomorrow seems to depend on building weapons which will destroy him tomorrow.

Frederick Alexander Lindemann [Lord Cherwell] 1886–1957

3   [To Lord De L'Isle (1957)] You know the definition of the perfectly designed machine . . . . The perfectly designed machine is one in which all its working parts wear out simultaneously. I am that machine.
in Lord Birkenhead *The Prof in Two Worlds* 1961 (London: Collins)

Eric Linklater [Robert Russell] 1899–1974

4   For the scientific acquisition of knowledge is almost as tedious as the routine acquisition of wealth.
White Man's Saga 1929 (London: Cape)

Carl Linnaeus 1707–1778

5   *Natura non facit saltus.*
Nature does not make jumps.
Philosophia Botanica 1751, no. 77

Gabriel Lippmann 1845–1921

6   [On the Gaussian curve, remarked to Poincaré] *Les expérimentateurs s'imaginent que c'est un théorème de mathématique, et les mathématiciens d'être un fait expérimental.*
Experimentalists think that it is a mathematical theorem while the mathematicians believe it to be an experimental fact.
in D'Arcy Thompson *On Growth and Form* 1917 (London: Cambridge UP) I

Konrad Lorenz 1903–

7   Historians will have to face the fact that natural selection determined the evolution of cultures in the same manner as it did that of species.
On Aggression 1966 (New York: Harcourt, Brace & World)

8   In nature we find not only that which is expedient, but also everything which is not so inexpedient as to endanger the existence of the species.
On Aggression 1966 (New York: Harcourt, Brace & World)

Ada Augusta [Countess of] Lovelace 1815–1853

1   The distinctive characteristic of the Analytical Engine, and that which has
rendered it possible to endow mechanism with such extensive faculties as
bid fair to make this engine the executive right-hand of abstract algebra, is
the introduction into it of the principle which Jacquard devised for regula-
ting, by means of punched cards, the most complicated patterns in the
fabrication of brocaded stuffs. It is in this that the distinction between
the two engines lies. Nothing of the sort exists in the Difference Engine.
We may say most aptly, that the Analytical Engine *weaves algebraical
patterns* just as the Jacquard-loom weaves flowers and leaves . . . . In
enabling mechanisms to combine together *general* symbols in successions of
unlimited variety and extent, a uniting link is established between the
operations of matter and the abstract mental processes of the most abstract
branch of mathematical science. A new, a vast, and a powerful language is
developed for the future use of analysis, in which to wield its truths so that
these may become of more speedy and accurate practical application for the
purposes of mankind than the means hitherto in our possession have
rendered possible . . .
*General Menabrea's Sketch of the Analytical Engine, invented by Charles Babbage. With
extensive notes by the translator* October 1842

Amy Lowell 1874–1925

2   Christ! What are patterns for?
Patterns in *The Complete Poetical Works of Amy Lowell* (Boston, Mass: Houghton Mifflin)

Lucian *ca* 115–*ca* 180

3   *Deus ex machina.*
A God from the machine.
[An allusion to the stage machinery of the theatre, for which see Mary Renault *The Mask of
Apollo*]

Lucretius 99–55 BC

4   I believe that this world is newly made: its origin is a recent event, not one
of remote antiquity. That is why even now some arts are still being per-
fected: the process of development is still going on. Many improvements
have just been introduced in ships. It is no time since organists gave birth
to their tuneful harmonies. Yes, and it is not long since the truth about
nature was first discovered, and I myself am even now the first who has
been found to render this revelation into my native speech.
De Rerum Natura

Georg Lukacs 1885–1971

5   Nature is a social category.
History and Class Consciousness transl R Livingstone, 1923 (London: The Merlin Press)

Graham Lusk

6   The work of a man's life is equal to the sum of all the influences he has
brought to bear upon the world in which he lives.
Science 1927 **65** 555

Martin Luther 1483–1546

1   *Die Arznei macht kranke, die Mathematik traurige und die Theologie sündhafte Leute.*
Medicine makes people ill, mathematics makes them sad and theology makes them sinful.

Trofim Denisovich Lysenko 1898–

2   The Party, the Government and J V Stalin personally, have taken an unflagging interest in the further development of the Michurin teaching.
[This was the session at which Mendelian genetics were condemned and outlawed for nearly a generation]
Report to the Lenin Academy of Agricultural Sciences Moscow, 31 July–7 August 1948

Thomas Babbington [Baron] Macaulay 1800–1859

3   The art which Bacon taught was the art of inventing arts.
Lord Bacon in *Edinburgh Review* July 1837

4   But even Archimedes was not free from the prevailing notion that geometry was degraded by being employed to produce anything useful. It was with difficulty that he was induced to stoop from speculation to practice. He was half ashamed of those inventions which were the wonder of hostile nations, and always spoke of them slightingly as mere amusements, as trifles in which a mathematician might be suffered to relax his mind after intense application to the higher parts of his science.
Lord Bacon in *Edinburgh Review* July 1837

Warren S McCulloch 1898–

5   Don't bite my finger—look where it's pointing.
[The pioneer of modelling the neurones and their operation]
in Stafford Beer *Platform for Change* 1975 (Chichester: Wiley)

Ernst Mach 1838–1916

6   The aim of research is the discovery of the equations which subsist between the elements of phenomena.
Popular Scientific Lectures Chicago, 1910

7   Every statement in physics has to state relations between observable quantities.
[Mach's Principle]

Antonio Machado 1875–1939

8   *Caminante, no hay camino*
Se hace camino al andar.
Traveller, there is no path,
Paths are made by walking.
[Theme of popular song in Latin America]

Alan Lindsay Mackay 1926–

9   How can we have any new ideas or fresh outlooks when 90 per cent of all

the scientists who have ever lived have still not died?
Scientific World 1969 **13** 17–21

1   Like the ski resort full of girls hunting for husbands and husbands hunting
for girls, the situation is not as symmetrical as it might seem.
Lecture, Birkbeck College, University of London, 1964

Charles Mackay 1814–1889

2   Blessings on Science, and her handmaid Steam
They make Utopia only half a dream.
Railways 1846

3   Cannon-balls may aid the truth,
But thought's a weapon stronger;
We'll win out battles by its aid;—
Wait a little longer.
The Good Time Coming

4   Truth . . . and if mine eyes
Can bear its blaze, and trace its symmetries,
Measure its distance, and its advent wait,
I am no prophet—I but calculate.
The Poetical Works of Charles Mackay 1876 (London: Frederick Warne)

[Sir] Halford John Mackinder 1861–1947

5   Knowledge is one. Its division into subjects is a concession to human
weakness.
Proceedings of the Royal Geographical Society 1887 **9** 141–60

Colin Maclaurin 1698–1746

6   [Writing to James Stirling in 1740] . . . an unlucky accident has happened
to the French mathematicians at Peru. It seems that they were shewing
some French gallantry to the native's wives, who have murdered their
servants, destroyed their Instruments and burn't their papers, the Gentlemen
escaping narrowly themselves. What an ugly article this will make in a
journal.
in Charles Tweedie *James Stirling* 1922 (Oxford: Oxford UP)

John McLeod and **John Osborn**

7   . . . in real life mistakes are likely to be irrevocable. Computer simulation,
however, makes it economically practical to make mistakes on purpose. If
you are astute, therefore, you can learn much more than they cost. Furthermore, if you are at all discreet, no one but you need ever know you made
a mistake.
Natural Automata and Useful Simulations ed H H Pattee *et al* 1966 (London: Macmillan)

Herbert Marshall McLuhan 1911–

8   When this circuit learns your job, what are you going to do?
The Medium is the Massage (New York: Basic Books)

Magna Charta 1215

1   There shall be standard measures of wine, beer, and corn—the London quarter—throughout the whole of our kingdom, and a standard width of dyed, russet and halberject cloth—two ells within the selvedges; and there shall be standard weights also.

Michael Maier 17th Century

2   *Fac ex mare et foemina circulum, inde quadrangulum, hinc triangulum, fac circulum et habebis lapidem philosophorum.*
From a man and a woman make a circle, then a square, then a triangle, finally a circle and you will obtain the Philosophers' Stone.
Scrutinium chymicum Frankfurt 1867

André Malraux 1901–1970

3   *Les intellectuels sont comme les femmes; ils s'en prennent aux militaires.*
Intellectuals are like women; they go for the military.
in N Moss *Men who play God* (London: Gollancz)

Thomas Robert Malthus 1766–1834

4   Above all, the Mechanics Institutions, open the fairest prospect that, within a moderate period of time, the fundamentals of political economy will, to a very useful extent, be known to the higher, middle, and a most important portion of the working classes of society in England.
An Essay on the Principles of Population 1798

5   Population, when unchecked, increases in a geometric ratio. Subsistence only increases in an arithmetic ratio.
An Essay on the Principles of Population 1798

6   The passion between the sexes . . . in every age . . . is so nearly the same that it may be considered in algebraic language as a given quantity.
An Essay on the Principles of Population 1798

Thomas Mann 1875–1955

7   A great truth is a truth whose opposite is also a great truth.
Essay on Freud 1937

8   I tell them that if they will occupy themselves with the study of mathematics they will find in it the best remedy against the lusts of the flesh.
The Magic Mountain 1927 (London: Secker)

9   Yet each in itself—this was the uncanny, the antiorganic, the life-denying character of them all—each of them was absolutely symmetrical, icily regular in form. They were too regular, as substance adapted to life never was to this degree—the living principle shuddered at this perfect precision, found it deathly, the very marrow of death—Hans Castorp felt he understood now the reason why the builders of antiquity purposely and secretly

introduced minute variations from absolute symmetry in their columnar structures.
The Magic Mountain 1927 (London: Secker)

Karl Mannheim 1893–1947

1   Even the categories in which experiences are subsumed, collected, and ordered vary according to the social position of the observer.
Ideology and Utopia 1936 (London: Routledge)

Mao Tse-tung 1893–

2   . . . . It is man's social being that determines his thinking. Once the correct ideas characteristic of the advanced class are grasped by the masses, these ideas turn into a material force which changes society and changes the world.
Probably *On Practice*

3   . . . hydrogen and oxygen aren't just transformed immediately in any old way into water. Water has its history too.
Mao Tse-tung unrehearsed ed S Schram, 1974 (London: Penguin)

4   The atomic bomb is a paper tiger . . . . Terrible to look at but not so strong as it seems.
in Anna Louise Strong *A World's Eye View from a Yenan Cave (An Interview with Mao Tse-tung)*

5   Dialectics was interpreted in the past as consisting of three big laws, and Stalin said that it consisted of four big laws. I think there is only one basic law, and that it is the law of contradiction. Quality and quantity, affirmation and negation, phenomenon and essence, content and form, necessity and freedom, possibility and reality etc, are all unity of opposites.
in Thomas G Hart *The Dynamics of Revolution* 1971, University of Stockholm

6   Knowledge is a matter of science, and no dishonesty or conceit whatsoever is permissible. What is required is definitely the reverse—honesty and modesty.
On Practice

7   Letting a hundred flowers blossom and a hundred schools of thought content is the policy for promoting the progress of the arts and the sciences and a flourishing socialist culture in our land. Different forms and styles in art should develop freely and different schools in science should contend freely. We think that it is harmful to the growth of art and science if administrative measures are used to impose one particular style of art or school of thought and to ban another. Questions of right and wrong in the arts and sciences should be settled through free discussion in artistic and scientific circles and through practical work in these fields. They should not be settled in summary fashion.
On the correct handling of contradictions among the people

8   Marxist philosophy holds that the most important problem does not lie in

understanding the laws of the objective world and thus being able to explain it, but in applying the knowledge of these laws actively to change the world.
On Practice

1   Natural science is one of man's weapons in his fight for freedom. For the purpose of attaining freedom in society, man must use social science to understand and change society and carry out social revolution. For the purpose of attaining freedom in the world of nature, man must use natural science to understand, conquer and change nature and thus attain freedom from nature.
Speech at the inaugural meeting of the Natural Science Research Society for the Border Regions

2   Where do correct ideas come from? Do they drop from the skies? No. Are they innate in the mind? No. They come from social practice, and from it alone; they come from three kinds of social practice, the struggle for production, the class struggle and scientific experiment.
Where Do Correct Ideas Come From?

Ramon Margalef 1919–

3   Probably the hypothesis holds everywhere that the less mature ecosystem feeds the more mature structures around it.
Perspectives in Ecological Theory 1968 (Chicago, Ill: University of Chicago Press)

Christopher Marlowe 1564–1593

4   I count religion but a childish toy,
And hold there is no sin but ignorance.
Machiavel in *The Jew of Malta* Prologue

5   Nowe therein of all Sciences (I speak still of humane and according to the humane conceits) is our Poet the Monarch. For he dooth not only show the way but giveth so sweet a prospect into the way as will entice any man to enter into it.
Apologie for Poetrie

Don Marquis 1878–1937

6   An idea isn't responsible for the people who believe in it.
The Sun Dial

Groucho Julius Marx 1895–

7   What has posterity ever done for me?
Attributed

Karl Marx 1818–1883

8   History itself is an actual part of natural history, of nature's development into man. Natural science will in time include the science of man as the science of man will include natural science: there will be one science.
Writings of the young Marx on Philosophy and Society ed L D Easton and K H Guddat, 1967 (New York: Doubleday)

1   Darwin's book is very important
     and serves me as a basis in natural
     science for the class struggle in
     history. One has to put up with the
     crude English method of develop-
     ment, of course. Despite all
     deficiencies not only is the death-
     blow dealt here for the first time to
     'teleology' in the natural sciences,
     but their rational meaning is
     empirically explained.
     Letter to Lasalle in *Marx–Angels Selected
     Correspondence, 1846–95* transl Dona Torr,
     1943 (London: Lawrence & Wishart)

2   The man who draws up a pro-
     gramme for the future is a reactionary.
     Letter to Beesley, UK

3   Mankind always takes up only such problems as it can solve; since, looking
     at the matter more closely, we will always find that the problem itself
     arises only when the material conditions necessary for its solution already
     exist or are at least in the process of formation.
     [God never sends mouths but sends meat (English Proverb, 1546), but we have few reasons for
     such confidence in this statement today]

4   Only the working class can . . . convert science from an instrument of class
     rule into a popular force . . . . Only in the Republic of Labour can science
     play its proper role.
     On the Paris Commune K Marx and F Engels (London: Lawrence & Wishart)

5   *Die Philosophen haben die Welt nur verschieden interpretiert, es kommt
     darauf an, sie zu verändern.* (Eleventh Thesis on Feuerbach.)
     The philosophers have only interpreted the world in various ways; the
     point, however, is to change it.
     [Epitaph on his tomb in Highgate Cemetery, London]

6   The product of mental labour—science—always stands far below its value,
     because the labour-time necessary to reproduce it has no relation at all to
     the labour-time required for its original production.
     Theories of Surplus Value

7   We know only a single science, the science of history. History can be
     contemplated from two sides, it can be divided into the history of nature
     and the history of mankind. However, the two sides are not to be divided
     off; as long as men exist the history of nature and the history of men are
     mutually conditioned.
     The German Ideology

John Masefield 1878–1967

8   . . . Harwell Man's perpetual treasure-trove . . .
     in *The Collected Poems of John Masefield* 1923 (London: Heinemann)

Matsushita Electrical Company

1   For the building of a new Japan
Let's put mind and strength together,
Doing our best to promote production,
Sending our goods to the peoples of the world,
Endlessly and continuously,
Like water gushing from a fountain.
Grow industry, grow, grow, grow,
Harmony and sincerity. Matsushita Electrical.
[Company hymn. Sung like a school song, on official occasions, to promote *esprit de corps*]
in F L K Hsu *Iemoto, The Heart of Japan* 1975 (New York: Wiley)

William Somerset Maugham 1874–1965

2   It is bad enough to know the past; it would be intolerable to know the future.
in Richard Hughes *Foreign Devil* 1972 (London: Deutsch)

James Clerk Maxwell 1813–1879

3   ... that, in a few years, all great physical constants will have been approximately estimated, and that the only occupation which will be left to men of science will be to carry these measurements to another place of decimals.
[Maxwell himself categorically rejected this view and was attacking it]
October 1871. *Scientific Papers* **2** 244

4   In the very beginning of science,
the parsons, who managed things then,
Being handy with hammer and chisel,
made gods in the likeness of men;
Till Commerce arose and at length
some men of exceptional power
Supplanted both demons and gods by
the atoms, which last to this hour.
[Plus further verses]
Notes of the President's Address, 1874

5   The only laws of matter are those which our minds must fabricate, and the only laws of mind are fabricated for it by matter.
in J G Crowther *British Scientists of the Nineteenth Century* (London: Secker & Warburg)

[Sir] Peter Brian Medawar 1915–

6   Considered in its entirety, psychoanalysis won't do. It is an end product, moreover, like a dinosaur or a zeppelin; no better theory can ever be erected on its ruins, which will remain for ever one of the saddest and strangest of all landmarks in the history of twentieth century thought.
The Hope of Progress 1972 (London: Methuen)

7   No scientist is admired for failing in the attempt to solve problems that lie beyond his competence. The most he can hope for is the kindly contempt

earned by the Utopian politician. If politics is the art of the possible, research is surely the art of the soluble. Both are immensely practical-minded affairs. Good scientists study the most important problems they think they can solve. It is, after all, their professional business to solve problems, not merely to grapple with them.
The Art of the Soluble 1967 (London: Methuen)

[On Teilhard de Chardin *The Phenomenon of Man*] The greater part of it, I shall show, is nonsense, tricked out with a variety of tedious metaphysical conceits, and its author can be excused of dishonesty only on the grounds that before deceiving others he has taken great pains to deceive himself.
The Art of the Soluble 1967 (London: Methuen)

Charles Edward Kenneth Mees 1882–1960

2   The best person to decide what research shall be done is the man who is doing the research. The next best is the head of the department. After that you leave the field of best persons and meet increasingly worse groups. The first of these is the research director, who is probably wrong more than half the time. Then comes a committee which is wrong most of the time. Finally there is a committee of company vice-presidents, which is wrong all the time.
[Uttered in 1935. Mees was Research Director of Kodak Ltd]
Biographical Memoirs of Fellows of the Royal Society 1961 **7** 182

Herman Melville 1819–1891

3   Captain Ahab: 'My means are sane, my motive and my object mad'.
Moby Dick 1851

Dmitri Ivanovich Mendeleev 1834–1907

4   I am not afraid of the admission of foreign, even of socialistic ideas into Russia, because I have faith in the Russian people who have already got rid of the Tatar domination and the feudal system.

Robert K Merton 1910–

5   Most institutions demand unqualified faith; but the institution of science makes skepticism a virtue.
Social Theory and Social Structure 1962 (New York: Free Press)

[Prince] Clemens Wenzel Lothar Metternich-Winneburg 1773–1859

6   [To the Austrian Ambassador in London for transmission to King George IV, 1825] There is one matter which I beg you to bring to the King's notice yet again before your departure; this is the proposed foundation of a university of London. You have my authority to tell His Majesty of my absolute conviction that the implementation of this plan would bring about England's ruin.
G de Bertier De Sauvingny *Metternich and his times* in *University of London Bulletin* no. 2, January 1972

Ivan Vladimirovich Michurin 1855–1935

1   We must not wait for favours from Nature; our task is to wrest them from her.

[Slogan of the Lysenkoist school]
Short Dictionary of Philosophy Moscow 1955

Ludwig Mies van der Rohe 1886–1969

2   Less is more.

[The architect affirms the positive virtue of shaving with Occam's Razor (q.v.)]
Obituary in *The Times* 19 August 1969

3   The long path from material through function to creative work has only one goal: to create order out of the desperate confusion of our time.

John Stuart Mill 1806–1873

4   The habit of analysis has a tendency to wear away the feelings.

Autobiography v

Henry Miller 1891–

5   The wallpaper with which the men of science have covered the world of reality is falling to tatters.

The Tropic of Cancer 1934 (London: Calder)

John Milton 1608–1674

6   *The Argument:* Adam inquires concerning celestial motions; is doubtfully answered, and exhorted to search rather things more worthy of knowledge.

[Kepler died in 1630]
Heading in *Paradise Lost* 1667, Book VIII

7   Behold now this vast City: a city of refuge, the mansion house of liberty, encompassed and surrounded with His protection; the shop of war hath not there more anvils and hammers waking, to fashion out plates and instruments of armed justice in defence of beleaguered Truth, than there be pens and hands there, sitting by their studious lamps, musing, searching, revolving new notions . . .

[Describing London during the Civil War]
Areopagitica

8   Eccentric, intervolved, yet regular
Then most, when most irregular they seem;
And in their motions harmony divine.

Paradise Lost 1667, Book V, 623

9   From Man or Angel the great Architect
Did wisely to conceal, and not divulge,
His secrets, to be scanned by them who ought
Rather admire. Or, if they list to try
Conjecture, he his fabric of the Heavens

Hath left to their disputes—perhaps to move
His laughter at their quaint opinions wide
Hereafter, when they come to model Heaven
And calculate the stars: how they will wield
The mighty frame: how build, unbuild, contrive
To save appearances; how gird the Sphere
With Centric and Eccentric scribbled o'er,
Cycle and Epicycle, Orb in Orb.
Paradise Lost 1667, Book VIII, 72

1   [Mulciber, the architect of the great palace of Pandemonium, had been thrown out of Heaven]
. . . nor aught availed him now
To have built in Heaven high towers; nor did he 'scape
By all his engines, but was headlong sent
With his industrious crew to build in Hell.
[The theory of the takeover which ousted Satan and his colleagues is entertainingly discussed by Anthony Jay in *Management and Machiavelli* 1970 (London: Penguin)]
Paradise Lost 1667, Book I, 748

2   [The Tree of Knowledge] O Sacred, Wise and Wisdom-giving Plant, Mother of Science.
Paradise Lost 1667 Book IX, 679

Alwyn Mittasch 1869–1953

3   Chemistry without catalysis, would be a sword without a handle, a light without brilliance, a bell without sound.
Journal of Chemical Education 1948 531–2

[Sir] Walter Hamilton Moberly 1881–

4   For God's sake, stop researching for a while and begin to think.
The Crisis in the University 1949 (London: Student Christian Movement Press)

J Moleschott 1822–1893

5   [Life is] woven out of air by light.
Perspectives in Biology and Medicine 1972, Winter, p208

Michel Eyquem Montaigne 1533–1592

6   If, by being overstudious, we impair our health and spoil our good humour . . . let us give it over.
Essais

7   Science without conscience is but death of the soul.
Essais

[Sir] Thomas More 1478–1535

8   Herodicus, being a trainer, and himself of a sickly constitution, by a combination of training and doctoring found out a way of torturing first and

chiefly himself, and secondly the rest of the world. By the invention of lingering death; for he had a mortal disease; which he perpetually tended, and, as recovery was out of the question, he passed his entire life as a valetudinarian; he could do nothing but attend upon himself, and he was in constant torment whenever he departed in anything from his usual regimen, and so dying hard, by the help of science he struggled on to old age.

Utopia transl P K Marshall, 1965 (New York: Washington Square Press)

Augustus De Morgan 1808–1871

1   Lagrange, in one of the later years of his life, imagined that he had overcome the difficulty (of the parallel axiom). He went so far as to write a paper, which he took with him to the Institute, and began to read it. But in the first paragraph something struck him which he had not observed: he muttered: '*Il faut que j'y songe encore*', and put the paper in his pocket. [I must think about it again]

Budget of Paradoxes London, 1872

Christopher Morley 1890–

2   An engineer gave me an ashtray
Made of a chunk of smelted bismuth.
The ore, when cooked,
Crystallises in sharp stairs and corners,
Like the ruins of a mimic Cuzco.
   O basic and everlasting geometry.
The cordillera itself
In the slack and purge of the fire
Boils into right angles,
Takes conventional Inca pattern.
The greatest disorder on earth
Has the instinct of Perfect Form.

[Lord] John Morley [of Blackburn] 1838–1923

3   The next great task of science is to create a religion for mankind.

Samuel Finley Breese Morse 1791–1872

4   What hath God wrought.

[First message sent by him over the electric telegraph, 24 May 1844]

Herbert Joseph Muller 1905–

5   Another way of describing the revolution in physics is to say that the key nouns have been changed into verbs—to move, to act, to happen. What moves and acts, physicists do not care; 'matter' to them means 'to matter', to make a difference. But our language is still geared to express 'states of being', rather than processes. In this connection, also, the German language helps to explain German philosophy. The Germans have been especially prone to hypostatize their abstractions, identify the Rational and the Real, invent concepts comparable to Frankfurterness and Sauerkrautitude—for

they capitalize all their nouns. And this may help to explain their present worship of the State.
Science and Criticism 1943 (New Haven, Conn: Yale UP)

1   The great revolutionary thinkers are those who most violently wrenched traditional associations: Karl Marx was a philosophical Oscar Wilde, more scandalous because more sober.
Science and Criticism 1943 (New Haven, Conn: Yale UP)

Hermann Joseph Muller 1890–1967

2   Death is an advantage to life . . . . Its advantage lies chiefly in giving ampler opportunities for the genes of the new generation to have their merits tested out . . . by clearing the way for fresh starts . . .
Science 1955 **121** 1

Fridtjof Nansen 1861–1930

3   Man wants to know, and when he ceases to do so, he is no longer man.
[On the reason for polar explorations]

Napoleon Bonaparte 1769–1821

4   The advance and perfecting of mathematics are closely joined to the prosperity of the nation.

5   They may say what they like; everything is organized matter.

Gamel Abdel Nasser 1918–1970

6   The genius of you Americans is that you never make clear-cut stupid moves, only complicated stupid moves which make us wonder at the possibility that there may be something to them [which] we are missing.
[A text book of the game-theory view of politics as practiced by the CIA]
in Miles Copeland *The Game of Nations* 1969 (London: Weidenfeld & Nicolson)

Joseph Needham 1900–

7   But Chinese civilization has the overpowering beauty of the wholly other, and only the wholly other can inspire the deepest love and the profoundest desire to learn.
The Grand Titration 1969 (London: Allen & Unwin)

8   Democracy might therefore almost in a sense be termed that practice of which science is the theory.
The Grand Titration 1969 (London: Allen & Unwin)

9   *Laboratorium est oratorium.*
The place where we do our scientific work is a place of prayer.

Henry Needler 1685–1760

10   Who formed the curious texture of the eye,

And cloath'd it with the various tunicles,
And texture exquisite; with chrystal juice
Supply'd it, to transmit the rays of light?

A Poem to Prove the Certainty of a God in *Miscellaneous Correspondence* ed Benjamin Martin, London, 1759

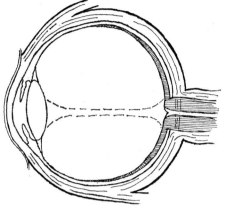

Jawaharlal Nehru 1889–1964

1   I fear that the spinning wheel is not stronger than the machine.

[Gandhi and his followers promoted village industry rather than industrialization and their spinning wheel appears on the flag of India]

2   It is science alone that can solve the problems of hunger and poverty, in-sanitation and illiteracy, of superstition and deadening custom and tradition, of vast resources running to waste, of a rich country inhabited by starving people . . . . Who indeed could afford to ignore science today? At every turn we have to seek its aid . . . . The future belongs to science and to those who make friends with science.

Proceedings of the National Institute of Sciences of India 1961 **27A** 564

[Cardinal] John Henry Newman 1801–1890

3   Living movements do not come out of committees.

[Sir] Isaac Newton 1642–1727

4   Are not gross bodies and light convertible into one another; and may not bodies receive much of their activity from the particles of light which enter into their composition? The changing of bodies into light, and light into bodies, is very comfortable to the course of Nature, which seems delighted with transmutations.

Opticks 1704, Query 30

5   I feign no hypotheses (hypotheses *non fingo*), for whatever is not deduced from the phenomena is to be called a hypothesis, and hypotheses, whether metaphysical or physical, whether of occult qualities or mechanical have no

place in experimental philosophy. In this philosophy particular propositions are inferred from the phenomena and afterwards rendered general by induction.

Scholium to *Philosophiae Naturalis Principia Mathematica*

1   I intend, to be no further solicitous about matters of Philosophy; and therefore I hope you will not take it ill, if you find me never doing anything more in that kind.

[Letter to Oldenburg, the Secretary of the Royal Society]
Opticks 1704

2   I know not what I may appear to the world, but to myself I seem to have been only like a boy playing on the sea-shore, and diverting myself in now and then finding a smoother pebble or a prettier shell than ordinary, whilst the great ocean of truth lay all undiscovered before me.

in D Brewster *Memoirs of Newton* 1855, vol 2, ch 27

3   Physics, beware of metaphysics.

4   Whence is it that nature does nothing in vain; and whence arises all that order and beauty which we see in the world?

Opticks 1704

Norman Nicholson 1914–

5   The toadstool towers infest the shore:
Stink-horns that propagate and spore
Wherever the wind blows.
Scafell looks down from the bracken band,
And sees hell in a grain of sand,
And feels the canker itch between his toes.
     This is a land where dirt is clean,
And poison pasture, quick and green,
And storm sky, bright and bare;
Where sewers flow with milk, and meat
Is carved up for the fire to eat,
And children suffocate in God's fresh air.

[On the leak of radioactive iodine from Windscale]
Windscale in *A Local Habitation* 1972 (London: Faber & Faber)

6   The furthest stars recede
Faster than the earth itself to our need.
     For far beyond the furthest, where
Light is snatched backward, no
Star leaves echo or shadow
To prove it has ever been there.
     And if the universe
Reversed and showed
The colour of its money;
If now unobservable light
Flowed inward, and the skies snowed
A blizzard of galaxies,

The lens of night would burn
Brighter than the focussed sun,
And man turn blinded
With white-hot darkness in his eyes.

The Expanding Universe in *The Pot Geranium* 1954 (London: Faber & Faber)

Marjorie Hope Nicolson 1894–

1   The language of poetry and science was no longer one when the world was no longer one.

in *Encyclopaedia of Poetry & Poetics* 1965 (Princeton, NJ: Princeton UP)

Friedrich Nietzsche 1844–1900

2   . . .—it is all over with priests and gods when man becomes scientific. Moral: science is the forbidden as such—it alone is forbidden. Science is the first sin, seed of all sin, the original sin. This alone is morality. 'Thou shalt not know'—the rest follows.

Antichrist ch 8

3   *Glaubt ihr denn, dass die Wissenschaften entstanden und gross geworden wären, wenn ihnen nicht Zauberer, Alchimisten, Astrologen and Hexen vorangelaufen wären als die, welche erst Durst, Hunger und Wohlgeschmack an verborgenen und verbotenen Mächten schaffen mussten?*
Do you believe then that the sciences would ever have arisen and become great if there had not beforehand been magicians, alchemists, astrologers and wizards, who thirsted and hungered after abscondite and forbidden powers?

Die fröhliche Wissenschaft 1886, IV

Florence Nightingale 1820–1910

4   [Of her] Her statistics were more than a study, they were indeed her religion. For her Quetelet was the hero as scientist, and the presentation copy of his *Physique sociale* is annotated by her on every page. Florence Nightingale believed—and in all the actions of her life acted upon that belief—that the administrator could only be successful if he were guided by statistical knowledge. The legislator—to say nothing of the politician—too often failed for want of this knowledge. Nay, she went further; she held that the universe—including human communities—was evolving in accordance with a divine plan; that it was man's business to endeavour to understand this plan and guide his actions in sympathy with it. But to understand God's thoughts, she held we must study statistics, for these are the measure of His purpose. Thus the study of statistics was for her a religious duty.

in Karl Pearson *The life, letters and labours of Francis Galton* vol 2, 1924 (London: Cambridge UP). *Isis* **8** 186

Novalis [Friedrich von Hardenberg] 1772–1801

5   *Die Mathematik ist das Leben der Götter.*
Mathematics is the life of the gods.

Charles Kay Ogden 1889–1957

1   The belief that words have a meaning of their own account is a relic of primitive word magic, and it is still a part of the air we breathe in nearly every discussion.
The Meaning of Meaning 1923 (London: Kegan Paul)

Charles Kay Ogden and Ivor Armstrong Richards 1889–1957 and 1893–

2   The gostak distims the doshes.
[There is an excellent science-fiction story on this theme]
The Meaning of Meaning 1923 (London: Kegan Paul)

Henry Oldenburg *ca* 1626–1678

3   I acknowledge that the jealousy about the first authors of experiments which you speak of, is not groundless; and therefore offer myself to register all those you, or any person, shall please to communicate as new, with that fidelity, which both the honour of my relation to the Royal Society (which is highly concerned in such experiments) and my own inclinations, do strongly oblige me to.
[The first editor of the *Philosophical Transactions of the Royal Society* writing to Robert Boyle]
Isis 1940 **31** 321

Bernard More Oliver 1916–

4   It is time that science, having destroyed the religious basis for morality, accepted the obligation to provide a new and rational basis for human behaviour—a code of ethics concerned with man's needs on earth, not his rewards in heaven.
Towards a New Morality in *IEEE Spectrum* 1972 **9** 52

Omar Khayyam *ca* 1050–*ca* 1123

5   Myself when young did eagerly frequent
Doctor and Saint, and heard great Arguement
About it and about: but everymore
Came out by the same door as in I went.
The Rubiyat 1859, transl Edward Fitzgerald, l.28

Julius Robert Oppenheimer 1904–1967

6   In some sort of crude sense which no vulgarity, no humor, no overstatement can quite extinguish, the physicists have known sin; and this is a knowledge which they cannot lose.
[Lecture at MIT 25 November 1947]
Physics in the Contemporary World (Cambridge, Mass: Harvard UP)

7   There are children playing in the street who could solve some of my top problems in physics, because they have modes of sensory perception that I lost long ago.

8   There floated through my mind a line from the *Bhagavad-Gita* in which Krishna is trying to persuade the Prince that he should do his duty: 'I am

become death, the shatterer of worlds'. I think we all had this feeling more or less.

[On 16 July 1945, at the test of the first atomic bomb—Trinity. The previous lines of the *Bhagavad-Gita* are:
If the radiance of a thousand suns
Were to burst into the sky,
that would be like
the splendour of the Mighty One]
in N P Davis *Lawrence and Oppenheimer* 1969 (London: Cape)

1   When you see something that is technically sweet, you go ahead and do it and you argue about what to do about it only after you have had your technical success.

in R W Reid *Tongues of Conscience: Weapons Research and the Scientist's Dilemma* 1969 (New York: Walker)

Martin Oppenheimer 1930–

2   Today's city is the most vulnerable social structure ever conceived by man.

Urban Guerilla (London: Quadrangle)

José Ortega y Gasset 1883–1955

3   Contemporary science, with its system and methods, can put blockheads (*tontos*) to good use.

Obras Completas. Revista de Occidente 1958 **6** 143

George Orwell 1903–1950

4   In a way it is even humiliating to watch . . . miners working. It raises in you a momentary doubt about your own status as an 'intellectual' and a superior person generally. For it is brought home to you, at least while you are watching, that it is only because miners sweat their guts out that superior persons can remain superior. You and I and the editors of the *Times Literary Supplement*, and the Poets and the Archbishop of Canterbury and Comrade X, author of *Marxism for Infants*—all of us really owe the comparative decency of our lives to poor drudges underground . . . with their throats full of . . . dust, driving their shovels forward with arms and belly muscles of steel.

Down the Mines in *The Road to Wigan Pier* 1937 (London: Secker & Warburg)

5   Who controls the past controls the future. Who controls the present controls the past.

Nineteen eighty-four 1949 (London: Secker & Warburg)

[Sir] William Osler 1849–1919

6   In science the credit goes to the man who convinces the world, not to the man to whom the idea first occurs.

Probably *Aequanimitas* with other Addresses in *Books and Men*

Ovid 43 BC–AD 17

7   *Nihil est toto, quo perstet, in orbe*
Cuncta fluunt, omnisque vagans formatur imago
Ipsa quoque odsidue labuntur tempora motu.

There is nothing in the whole world which is permanent
Everything flows onward; all things are brought into being with a changing
nature;
The ages themselves glide by in constant movement.
Metamorphoses XV, i, 177

Axel Gustafsson [Count] Oxenstierna 1583–1654

1   *Nescis, mi fili, quantilla ratione mundus regatur.*
     You don't know, my dear boy, with what little reason the world is
     governed.

[Sir] James Paget 1814–1899

2   You will find that fatigue has a larger share in the promotion and trans-
     mission of disease than any other single condition you can name.
     Science in War 1940 (London: Penguin)

Bernard Palissy 1510–1589

3   You must know that, in order to manage well a kiln full of pottery, even
     when it is glazed, you must control the fire by so careful a philosophy that
     there would be no spirit however noble which would not be much tried
     and often disappointed. As to the manner of filling your kiln, a singular
     geometry is needed . . . . The arts for which compass, ruler, numbers,
     weights and measures are needed should not be called mechanics.
     [Palissy was the potter who found science for himself]
     L'Art de Terre in *Oeuvres Completes* Paris, 1884

George Paloczi-Horvath 1908–1973

4   Students of Soviet affairs know how difficult it is to foretell the Soviet past.
     Khrushchev 1960 (Boston, Mass: Little, Brown)

Paracelsus [Philippus Aureolus Theophrastus Bombastus von Hohenheim] 1493–1541

5   To each elemental being the element in which it lives is transparent,
     invisible and respirable, as the atmosphere to ourselves.
     F Hartmann *The Life of . . . Paracelsus* 1896, London, 2nd edn

6   What is accomplished with fire is alchemy, whether in the furnace or the
     kitchen stove.
     in J Bronowski *The Ascent of Man* 1975 (London: BBC)

Vilfredo Pareto 1848–1923

7   Give me fruitful error any time, full of seeds, bursting with its own
     corrections. You can keep your sterile truth for yourself.
     [Comment on Kepler]

8   In a dispute between two chemists there is a judge; Experience. In a
     dispute between a Moslem and a Christian, who is the judge? Nobody.
     The Mind and Society

[Sir] Alan Sterling Parkes 1900–

1 . . . no woman should be kept on the Pill for 20 years until, in fact, a sufficient number have been kept on the Pill for 20 years.
Nature 1970 **226** 187

Cyril Northcote Parkinson 1909–

2 Work expands so as to fill the time available for its completion (Parkinson's First Law).
Parkinson's Law 1957 (London: Murray)

3 Expenditure rises to meet income (Parkinson's Second Law).
In Laws and Outlaws 1962

4 Expansion means complexity, and complexity decay (Parkinson's Third Law).
In Laws and Outlaws 1962

5 The Law of Triviality:
Briefly stated, it means that the time spent on any item of the agenda will be in inverse proportion to the sum involved.
Parkinson's Law 1957 (London: Murray)

Talcott Parsons 1902–

6 Science is intimately integrated with the whole social structure and cultural tradition. They mutually support one another—only in certain types of society can science flourish, and conversely without a continuous and healthy development and application of science such a society cannot function properly.
The Social System 1951 (New York: Free Press) ch VIII

Blaise Pascal 1623–1662

7 [I feel] engulfed in the infinite immensity of spaces whereof I know nothing, and which know nothing of me, I am terrified . . . . The eternal silence of these infinite spaces alarms me.
Pensées 1657

Louis Pasteur 1822–1895

8 *Dans les champs de l'observation, l'hasard ne favorise que les esprits préparés.*
In the field of observation, chance only favours those minds which have been prepared.
Encyclopaedia Britannica 1911, 11th edn, vol 20

9 *Il faut de toute necessité que des actions dissymetriques president pendant la vie à l'élaboration des vrais principes immédiats naturels dissymetriques. Quelle peut être la nature de ces actions dissymetriques? Je pense, quant à moi, qu'elles sont d'ordre cosmique. L'univers est un énsemble dissymetrique et je suis persuadé que la vie, telle qu'elle se manifeste à nous, est fonction de la dissymetrie de l'univers ou des consequences qu'elle entraîne. L'univers est dissymetrique.*

It is inescapable that asymmetric forces must be operative during the synthesis of the first asymmetric natural products. What might these forces be? I, for my part, think that they are cosmological. The universe is asymmetric and I am persuaded that life, as it is known to us, is a direct result of the asymmetry of the universe or of its indirect consequences. The universe is asymmetric.

Comptes Rendus de l'Académie des Sciences 1 June 1874. Reprinted *Oeuvres* 1, 361. In J B S Haldane *Nature* 1960 **185** 87

1   There does not exist a category of science to which one can give the name applied science. There are science and the applications of science, bound together as the fruit of the tree which bears it.

Pourquoi la France n'a pas trouvé d'hommes supérieurs au moment du peril in *Revue Scientifique* 1871

Wolfgang Pauli 1900–1958

2   *Ich habe nichts dagegen wenn Sie langsam denken, Herr Doktor, aber ich habe etwas dagegen wenn Sie rascher publizieren als denken.*
I don't mind your thinking slowly: I mind your publishing faster than you think.

Attributed

Ivan Petrovich Pavlov 1849–1936

3   What can I wish to the youth of my country who devote themselves to science? . . . Thirdly, passion. Remember that science demands from a man all his life. If you had two lives that would not be enough for you. Be passionate in your work and in your searching.

Bequest to Academic Youth 1936

Karl Pearson 1857–1936

4   Modern science, as training the mind to an exact and impartial analysis of facts, is an education specially fitted to promote citizenship.

The Grammar of Science 1911 (London: A & C Black)

5   The right to live does not connote the right of each man to reproduce his kind . . . . As we lessen the stringency of natural selection, and more and more of the weaklings and the unfit survive, we must increase the standard, mental and physical, of parentage.

Darwinism, Medical Progress and Parentage 1912, University of London, University College Eugenics Laboratory, 2nd edn

6   The unity of all science consists alone in its method, not in its material.

The Grammar of Science 1911 (London: A & C Black)

Benjamin Peirce 1809–1880

7   Mathematics is the science which draws necessary conclusions.

[Memoir read before the National Academy of Sciences in Washington, 1870]
American Journal of Mathematics 1881 **4** 97

Pelagius [Morgan] *ca* 360–*ca* 420

8   *Si necessitatis est, peccatum non est; si voluntatis, vitari potest.*

If it is a necessity, then it is not a sin; if it is optional, then it can be avoided.
[The founder of the Pelagian heresy]

Pericles 5th Century BC

1   Although only a few may originate a policy, we are all able to judge it.
Funeral Oration 431 BC. *Thucydides* II, 41

Laurence Johnston Peter 1919–

2   The Peter Principle:
In a hierarchy every employee tends to rise to his level of incompetence.
[Whence two sub-principles]—In time, every post tends to be occupied by an employee who is incompetent to carry out its duties.—Work is accomplished by those employees who have not yet reached their level of incompetence.
The Peter Principle 1969 (New York: Morrow)

[Sir] William Petty 1623–1687

3   Nor do I doubt if the most formidable armies ever heere upon earth is a sort of soldiers who for their smallness are not visible.
[On microbes, 1640]
The Petty Papers 1927 (London: Constable)

Physical Review Letters

4   Scientific discoveries are not the proper subject for newspaper scoops and all media of mass communication should have equal opportunity for simultaneous access to the information. In the future we may reject papers whose main contents have been published previously in the daily press.
[Editorial by Professor Samuel A Goudsmit, 1 January 1960. Since Professor Goudsmit's retirement as editor 'the publicity-hungry high-energy physicists have succeeded in getting this old policy rescinded']

Jean Piaget 1896–

5   In short, the notion of structure is comprised of three key ideas: the idea of wholeness, the idea of transformation, and the idea of self-regulation.
Structuralism transl C Maschler, 1971 (London: Routledge & Kegan Paul)

Pablo [Ruiz y] Picasso 1881–1973

6   The against comes before the for.

7   *Je ne cherche pas; je trouve.*
I do not search; I find.
Étude de femme in P Oster *Nouveau Dictionnaire de Citations Françaises* 1970 (Paris: Librairie Hachette, Tchou Editeurs)

Charles Santiago Sanders Pierce 1839–1914

8   The essence of belief is the establishment of a habit.
Illustrations of the Logic of Science, II in *Popular Science Monthly, NY* January 1878

[Sir] Alfred Brian Pippard 1920–

9   The value of a formalism lies not only in the range of problems to which

it can be successfully applied, but equally in the degree to which it encourages physical intuition in guessing the solution of intractable problems.
Physics Bulletin 1969 **20** 455

Robert Pirsig 1929–

1   The way to solve the conflict between human values and technological needs is not to run away from technology, that's impossible. The way to resolve the conflict is to break down the barriers of dualistic thought that prevent a real understanding of what technology is—not an exploitation of nature, but a fusion of nature and the human spirit into a new kind of creation that transcends both.
Zen and the Art of Motorcycle Maintenance 1974 (London: Bodley Head)

2   The Buddha, the Godhead, resides quite as comfortably in the circuits of a digital computer or the gears of a cycle transmission as he does at the top of a mountain or in the petals of a flower.
Zen and the Art of Motorcycle Maintenance 1974 (London: Bodley Head)

3   A motorcycle functions entirely in accordance with the laws of reason, and a study of the art of motorcycle maintenance is really a miniature study of the art of rationality itself.
Zen and the Art of Motorcycle Maintenance 1974 (London: Bodley Head)

[Pope] Pius IX 1792–1878

4   Syllabus of the principal errors of our time . . . 12. The decrees of the Apostolic See and of the Roman congregation impede the true progress of science.
[21 December 1863]
in *Dogmatic Canons and Decrees* 1912 (Old Greenwich, Conn: Devin–Adair)

[Pope] Pius XII 1876–1958

5   The Church welcomes technological progress and receives it with love, for it is an indubitable fact that technological progress comes from God and, therefore, can and must lead to Him.
[Christmas Message, 1953]

6   One Galileo in two thousand years is enough.
[On being asked to proscribe the works of Teilhard de Chardin]
Attributed. See Stafford Beer *Platform for Change* 1975 (Chichester: Wiley)

Max Planck 1858–1947

7   An important scientific innovation rarely makes its way by gradually winning over and converting its opponents: it rarely happens that Saul becomes Paul. What does happen is that its opponents gradually die out, and that the growing generation is familiarised with the ideas from the beginning.
in G Holton *Thematic Origins of Scientific Thought* 1973 (Cambridge, Mass: Harvard UP)

Plato *ca* 429–347 BC

8   He who can properly define and divide is to be considered a god.
in Francis Bacon *Novum Organum* 1620, Book II, 26

1   The ludicrous state of solid geometry made me pass over this branch.
The Republic VII, 528

2   We must endeavour to persuade the principal men of our State to go and learn arithmetic, not as amateurs, . . . but for the sake of military use . . .
The Republic VII, 525

3   Let no one ignorant of geometry enter my door.
in John Tzetzes (*ca* 1110–*ca* 1180) *Chiliad* 8, 972

4   He is unworthy of the name of man who is ignorant of the fact that the diagonal of a square is incommensurable with its side.
in Sophie Germain *Mémoire sur les surfaces élastiques*

5   *Socrates:* Shall we set down astronomy among the subjects of study?
Glaucon: I think so, to know something about the seasons, the months and the years is of use for military purposes, as well as for agriculture and for navigation.
Socrates: It amuses me to see how afraid you are, lest the common herd of people should accuse you of recommending useless studies.
The Republic VII, 527

6   *Theaetetus:* Science is sensation.
Socrates: You give an opinion that cannot be despised, since it was Protagoras's. Yet he expressed it in another way, by saying that man was the measure of all things.
Theaetetus 152

Georgi Valentinovich Plekhanov 1856–1918

7   Bourgeois scientists make sure that their theories are not dangerous to God or to capital.
Karl Marx

Plutarch *ca* 46–*ca* 127

8   But what most of all afflicted Marcellus was the death of Archimedes. For it chanced that he was by himself, working out some problem with the aid of a diagram, and having fixed his thoughts and his eyes as well upon the matter of his study, he was not aware of the incursion of the Romans, or of the capture of the city. Suddenly a soldier came upon him and ordered him to go with him to Marcellus. This Archimedes refused to do until he had worked out his problem and established his demonstration, whereupon the soldier flew into a passion, drew his sword, and dispatched him. However, it is generally agreed that Marcellus was afflicted at his death, and turned away from his slayer as from a polluted person, and sought out the kindred of Archimedes and paid them honour.
Lives transl J & W Langhorne, 1876 (London: Chatto)

9   Plato said that God geometrises continually.
Convivialum disputationum 8, 2

1   You know, of course, that Lycurgus expelled arithmetical proportion from Lacedaemon, because of its democratic and rabble-rousing character. He introduced geometric proportion, . . .
Moralia Loeb edn, vol IX

Po Chu-i 772–846

2   'Those who speak know nothing;
Those who know are silent'.
These words, as I am told,
Were spoken by Lao-Tze.
If we are to believe that Lao-Tze
Was himself one who knew,
How comes it that he wrote a book
Of five thousand words?
transl Arthur Waley *Chinese Poems* 1946 (London: Allen & Unwin)

Edgar Allan Poe 1809–1849

3   Science . . . hast thou not dragged Diana from her car?
Sonnet *To Science* ca 1827

Jules Henri Poincaré 1854–1912

4   *Les faits ne parlent pas.*
Facts do not speak.

5   Science is built up with facts, as a house is with stones. But a collection of facts is no more a science than a heap of stones is a house.
La Science et l'Hypothèse 1902 (Paris: Flammarion)

Michael Polanyi 1891–1976

6   The pursuit of science can be organized . . . in no other manner than by granting complete independence to all mature scientists. They will then distribute themselves over the whole field of possible discoveries, each applying his own special ability to the task that appears most profitable to him. The function of public authorities is not to plan research, but only to provide opportunities for its pursuit. All they have to do is to provide facilities for every good scientist to follow his own interest in science.
The Logic of Liberty 1951 (Chicago, Ill: University of Chicago Press)

7   The Republic of Science shows us an association of independent initiatives, combined towards an indeterminate achievement. It is disciplined and motivated by serving a traditional authority, but this authority is dynamic; its continued existence depends on its constant self-renewal through the originality of its followers.
    The Republic of Science is a Society of Explorers. Such a society strives towards an unknown future, which it believes to be accessible and worth achieving. In the case of scientists, the explorers strive towards a hidden reality, for the sake of intellectual satisfaction. And as they satisfy themselves, they enlighten all men and are thus helping society to fulfil its obligation towards intellectual self-improvement.
Minerva 1962 **1** 54–73

Marco Polo *ca* 1254–1324

1   Here is told of the city of Kinsai [Hangkow] . . . . In other streets, live harlots, of whom there are so many that I dare not say the number . . . . These women are very clever and expert in their allurements and endearments, and ever have appropriate words ready for every kind of person. So, when foreigners have once tasted of them, they remain, so to speak, beside themselves, and are so taken by their sweetness and charm, that they can never forget them. Thus it is that, when they return home, they say they have been in Kinsai, namely in the City of Heaven, and long to be able to return there. In yet other streets live all the leeches and all the astrologers, the latter of whom also teach reading and writing.
The Travels of Marco Polo ed L F Benedetto, transl A Ricci, 1931 (London: Routledge & Kegan Paul)

Polybius *ca* 204–*ca* 122 BC

2   Whenever it is possible to find out the cause of what is happening, one should not have recourse to the gods.
in K Von Fritz *The Theory of the Mixed Constitution in Antiquity* 1954 (New York: Columbia UP)

[Saint] Polycarp *ca* 69–*ca* 155

3   In all these monstrous demons is seen an art hostile to God.
[On the clepsydra]

Georges Pompidou 1911–1974

4   There are three roads to ruin; women, gambling and technicians. The most pleasant is with women, the quickest is with gambling, but the surest is with technicians.
Sunday Telegraph 26 May 1968

Alexander Pope 1688–1744

5   Epitaph on Newton:
Nature and Nature's laws lay hid in night:
God said, 'Let Newton be.' and all was light.
[Added by Sir John Collings Squire:
It did not last: the Devil shouting 'Ho.
Let Einstein be.' restored the status quo]
The Works of Alexander Pope 1871 (London: Murray)

6   For Forms of Government let fools contest;
What'er is best administered is best.
An Essay of Man III, line 303

7   How index-learning turns no student pale,
Yet holds the eel of science by the tail.
The Dunciad Book I, line 279

8   Lo, the poor Indian: whose untutored mind
Sees God in clouds, or hears him in the wind:
His soul proud science never taught to stray
Far as the Solar Walk or Milky Way.
An Essay on Man I, line 99

1   Not chaos-like together wash'd and bruis'd,
     But, as the world, harmoniously confus'd:
     Where order in variety we see,
     And where, though all things differ, all agree.
     Windsor Forest

2   One science only will one genius fit;
     So vast is art, so narrow human wit.
     An Essay on Criticism part I, line 60

3   Order is Heaven's first law.
     An Essay on Man IV, line 49

4   Those Rules of old discovered, not devised
     Are Nature still, but Nature methodized;
     Nature, like Liberty, is but restrained
     By the same Laws which first herself ordained.
     Perspectives in Biology and Medicine, 1971, Autumn, p105

5   Why has not man a microscopic eye?
     For this plain reason, man is not a fly.
     An Essay on Man I, line 193

[Sir] Karl Raimund Popper 1902–

6   But I shall certainly admit a system as empirical or scientific only if it is
     capable of being *tested* by experience. These considerations suggest that not
     the *verifiability* but the *falsifiability* of a system is to be taken as a criterion
     of demarcation. In other words: I shall not require of a scientific system
     that it shall be capable of being singled out, once and for all, in a positive
     sense: but I shall require that its logical form shall be such that it can be
     singled out, by means of empirical tests, in a negative sense: *it must be
     possible for an empirical scientific system to be refuted by experience.*
     The Logic of Scientific Discovery 1959 (London: Hutchinson)

7   Science is not a system of certain, or well-established, statements; nor is it
     a system which steadily advances towards a state of finality . . . . And our
     guesses are guided by the unscientific the metaphysical (though biologically
     explicable) faith in laws, in regularities which we can uncover—discover.
     Like Bacon, we might describe our own contemporary science—'the method
     of reasoning which men now ordinarily apply to nature'—as consisting of
     'anticipations, rash and premature' and as 'prejudices'.
     The Logic of Scientific Discovery 1959 (London: Hutchinson)

Ezra Pound 1885–1972

8   'You damn sadist,' said mr cummings,
     'you try to make people think.'
     Canto 89 in *The Cantos of Ezra Pound* 1956 (London: Faber & Faber and New York: New
     Directions) © Ezra Pound, 1956

9   Of all those young women not one has enquired the cause of the world
     Nor the modus of lunar eclipses.
     Homage to Sextus Propertius in *The Collected Shorter Poems of Ezra Pound* 1926 (London:
     Faber & Faber and New York: New Directions) © Ezra Pound, 1926

1   止 a gnomon,
Our science is from the watching of shadows . . .
Canto 85 in *The Cantos of Ezra Pound* 1956 (London: Faber & Faber and New York: New Directions) © Ezra Pound, 1956

Cedric Price 1934–

2   The reason for architecture is to encourage . . . people . . . to behave, mentally and physically, in ways they had previously thought impossible.
Exhibition, RIBA Heinz Gallery, 8 October 1975

Derek John de Solla Price 1922–

3   The disciplines which analyse science have been generated piecemeal, but show many signs of beginning to cohere into a whole which is greater than the sum of its parts. This new study might be called 'history, philosophy, sociology, psychology, economics, political science and operations research (etc) of science, technology, medicine (etc).
in *The Science of Science* ed M Goldsmith and A L Mackay, 1964 (London: Souvenir Press)

4   Science is not just the fruit of the tree of knowledge, it is the tree itself.
Lecture, London, 1964

1   Using any reasonable definition of a scientist, we can say that between 80 and 90 per cent of all the scientists that have ever lived are alive now. Now depending on what one measures and how, the crude size of science in manpower or in publications tends to double within a period of 10 to 15 years.
Little Science, Big Science 1963 (New York: Columbia UP)

Don Krasher Price 1910–

2   Science . . . cannot exist on the basis of a treaty of strict non-aggression with the rest of society; from either side, there is no defensible frontier.
Government and Science 1954 (New York: New York UP)

Joseph Priestley 1733–1804

3   It was ill policy in Leo the Tenth to patronise polite literature. He was cherishing an enemy in disguise. And the English hierarchy (if there be anything unsound in its constitution) has equal reason to tremble even at an air pump or an electrical machine.
Experiments and Observations on Different Kinds of Air 1775–1786

Proclus Diadochus 412–485

4   It is well known that the man who first made public the theory of irrationals perished in a shipwreck in order that the inexpressible and unimaginable should ever remain veiled. And so the guilty man, who fortuitously touched on and revealed this aspect of living things, was taken to the place where he began and there is for ever beaten by the waves.
[Attributed]
Scholium to Book X of *Euclid* V, 417

Protagoras *ca* 481–*ca* 411 BC

5   Man is the measure of all things, of things that are, that they are, of things that are not, that they are not.
in Diogenes Laertius *Vitae Philosophicus* IX, 51

6   Of the gods I know nothing, whether they exist or do not exist: nor what they are like in form. Many things stand in the way of knowledge—the obscurity of the subject, the brevity of human life.
in Diogenes Laertius *Vitae Philosophicus* IX, 51

Pierre Joseph Proudhon 1809–1865

7   [*De même qu'il y a*] *une science des phenomènes physiques qui ne repose que sur l'observation des faits, il doit exister aussi une science de la société, absolue, rigoureuse, basée sur la nature de l'homme et de ses facultés, et sur leurs rapports, science qu'il ne faut pas inventer mais découvrir.*
Inasmuch as there is a science of physical phenomena, which rests only on the observation of facts, there ought also to exist a science of society which should be absolute and rigorous and based on the nature of man, his faculties and their inter-relationships. This should be a science to be discovered, not invented.
L'Utilité de la Célébration du Dimanche 1839

Marcel Proust 1871–1922

1   Distances are only the relation of space to time and vary with that relation.
Cities of the Plain I, 3

François Quesnay 1694–1774

2   Commerce, like industry, is merely a branch of agriculture. It is agriculture which furnishes the material of industry and commerce and which pays both . . .
Grains in *Encyclopédie*

Adolphe Quetelet 1796–1874

3   'The average man'.
[The invention of the concept]

4   The more progress physical sciences make, the more they tend to enter the domain of mathematics, which is a kind of centre to which they all converge. We may even judge of the degree of perfection to which a science has arrived by the facility with which it may be submitted to calculation.
in E Mailly *Eulogy on Quetelet* 1874, Smithsonian Report

François Rabelais 1494–1553

5   *. . . questio subtillissima, utrum chimera in vacuo bombinans possit comedere secundas intentiones.*
. . . a most subtle question whether a chimaera bombinating in a vacuum can devour second intentions. (Oxford English Dictionary.)
Tiers Livre Book II, vii

6   Nature abhors a vacuum.
[Quoting the Latin proverb *natura vacuum abhorret*]
Gargantua Book 1, ch 5

7   Science without conscience is but the ruin of the soul.
Gargantua's letter to Pantagruel

Isador Isaac Rabi 1898–

8   There isn't a scientific community. It is a culture. It is a very undisciplined organisation.
1965. In D S Greenberg *The Politics of Pure Science* 1967 (New York: New American Library)
© D S Greenberg, 1967

[Sir] Walter Alexander Raleigh 1861–1922

9   In an examination those who do not wish to know ask questions of those who cannot tell.
Some Thoughts on Examinations

Srinivasa Ramanujan 1887–1920

10   G H Hardy: I remember once going to see him when he was lying ill at

Putney. I had ridden in taxi-cab No. 1729 and remarked that the number seemed to me rather a dull one, and that I hoped it was not an unfavourable omen. 'No,' he replied, 'it is a very interesting number; it is the smallest number expressible as the sum of two cubes in two different ways.'

in G H Hardy *Ramanujan* 1940 (London: Cambridge UP)

$$1^3+12^3 = 9^3+10^3$$

James Arthur Ramsay 1909–

1   The mammal is a highly-tuned physiological machine carrying out with superlative efficiency what the lower animals are content to muddle through with.

Anatol Rapoport 1911–

2   One cannot play chess if one becomes aware of the pieces as living souls and of the fact that the Whites and the Blacks have more in common with each other than with the players. Suddenly one loses all interest in who will be champion.

Strategy and Conscience 1964 (New York: Harper & Row)

Alastair Reid 1926–

3   'Counting': Ounce, dice, trice, quartz, quince, sago, serpent, oxygen, nitrogen, denim.

Ounce, dice, trice 1956 (Boston, Mass: Little, Brown)

Ivor Armstrong Richards 1893–

4   The properties of the instruments or apparatus employed enter into . . . belong with and confine the scope of the investigation.

Speculative Instruments in *Internal Colloquies* (London: Routledge & Kegan Paul)

Lewis Fry Richardson 1881–1953

5   Big[ger] whirls have little whirls,

That feed on their velocity;
And little whirls have lesser whirls,
And so on to viscosity.

[Summarizing his classic paper *The Supply of Energy from and to Atmospheric Eddies* (1920)]

Charles Robert Richet 1850–1935

1   I possess every good quality, but the one that distinguishes me above all is modesty.

[Nobel Laureat for medicine, 1913]
The Natural History of a Savant transl Oliver Lodge, 1927 (New York: Doran)

[Baron] Peter Ritchie-Calder 1906–

2   The academic and basic scientists are 'The Makers-Possible'; the applied scientists and the technologists are 'The Makers-to-Happen', and the technicians 'The Makers-to-Work'. And nowadays, with operations research, market research, quality control, etc, the commercial scientists might be called 'The Makers-to-Pay'.

The Evolution of Science 1963 (Paris: UNESCO/Mentor)

Jacques Rohault 17th Century

3   . . . it was by just such a hazard, as if a man should let fall a handful of sand upon a table and the particles of it should be so ranged that we could read distinctly on it a whole page of Virgil's *Aenead*.

Traité de Physique Paris, 1671. Transl 1723, II

Jules Romains 1885–

4   *Les gens bien portants sont des malades qui s'ignorent.*
Every man who feels well is a sick man neglecting himself.

Knock, ou le Triomphe de la Médecine 1923 (Paris: Gallimard)

[Sir] Ronald Ross 1857–1932

5   This day relenting God
Hath placed within my hand
A wondrous thing; and God
Be praised. At His command,
Seeking His secret deeds
With tears and toiling breath,
I find thy cunning seeds,
O million-murdering Death.
I know this little thing
A myriad men will save,
O Death where is thy sting?
Thy victory, O Grave?

[Describing his discovery of the life-cycle of the malaria parasite. 1897]
Poems by Ronald Ross 1928 (London: E Mathews & Marrot)

6   Now twenty years ago
This day we found the thing;
With science and with skill

We found; then came the sting—
What we with endless labour won
The thick world scorned;
Not worth a word today—
Not worth remembering.
[Ross received the Nobel Prize for medicine in 1902]
Written 20 August 1917

Jean Rostand 1894–

1   *Ou apprendre le métier de Dieu?*
Who can teach us God's business?

2   *Être adulte, c'est être seul.*
To be adult is to be alone.
Pensées d'un biologiste (Paris: Stock)

Theodore Roszak 1933–

3   Nature composes some of her loveliest poems for the microscope and
telescope.
Where the Wasteland Ends 1972 (London: Faber & Faber)

Joseph Roux 1834–1886

4   Science is for those who learn; poetry for those who know.
Meditations of a Parish Priest 1

Henry Augustus Rowland 1848–1901

5   He who makes two blades of grass grow where one grew before is the
benefactor of mankind, but he who obscurely worked to find the laws of
such growth is the intellectual superior as well as the greater benefactor of
mankind.
in D S Greenberg *The Politics of Pure Science* 1967 (New York: New American Library) © D S
Greenberg, 1967

Count Rumford [Benjamin Thompson] 1753–1814

6   It frequently happens that in the ordinary affairs and occupations of life,
opportunities present themselves of contemplating some of the most curious
operations of nature.
in H A Bent *The Second Law* 1965 (New York: Oxford UP)

John Ruskin 1819–1900

7   Lily: 'We looked at the books about crystals but they are so dreadful.'
The Ethics of the Dust, Ten Lectures to Little Housewives on the Elements of Crystallisation 1866
(London: Smith Elder)

8   May: 'Oh. Have the crystals faults like us?'
L: 'Certainly, May. Their best virtues are shown in fighting their faults.
And some have a great many faults; and some are very naughty crystals
indeed.'
The Ethics of the Dust, Ten Lectures to Little Housewives on the Elements of Crystallisation 1866
(London: Smith Elder)

[Lord] Bertrand Russell 1872–1970

1   . . . the general public has derived the impression that physics confirms practically the whole of the Book of Genesis. I do not myself think that the moral to be drawn from modern science is at all what the general public has thus been led to suppose. In the first place, the men of science have not said nearly as much as they are thought to have said, and in the second place what they have said in the way of support for traditional religious beliefs has been said by them not in their cautious, scientific capacity, but rather in their capacity of good citizens, anxious to defend virtue and property.
The Scientific Outlook 1931 (London: Allen & Unwin)

2   Animals studied by Americans rush about frantically, with an incredible display of hustle and pep, and at last achieve the desired result by chance. Animals observed by Germans sit still and think, and at last evolve the solution out of their inner consciousness.
[On Thorndike and Koehler]

3   Aristotle maintained that women have fewer teeth than men; although he was twice married, it never occurred to him to verify this statement by examining his wives' mouths.
The Impact of Science on Society 1952 (London: Allen & Unwin)

4   Can a society in which thought and technique are scientific persist for a long period, as, for example, ancient Egypt persisted, or does it necessarily contain within itself forces which must bring either decay or explosion . . .?
Lloyd Roberts Lecture *Can a Scientific Community be Stable?* to the Royal Society of Medicine, 29 November 1949

5   I am compelled to fear that science will be used to promote the power of dominant groups rather than to make men happy.
[Replying to J B S Haldane's optimistic view expressed in *Daedalus—Science and the Future*]
Icarus, or the Future of Science 1925 (London: Kegan Paul)

6   The number of a class is the class of all classes similar to a given class.
Principles of Mathematics 1903 (New York: Cambridge UP)

7   Pure mathematics is the class of all propositions of the form p implies q, where p and q are propositions containing one or more variables, the same in the two propositions, and neither p nor q contains any constants except logical constants . . . . The fact that all Mathematics is Symbolic Logic is one of the greatest discoveries of our age; and when this fact has been established, the remainder of the principles of mathematics consists in the analysis of Symbolic Logic itself.
Principles of Mathematics 1903 (New York: Cambridge UP)

8   The fundamental concept in social science is Power, in the same sense in which Energy is the fundamental concept in physics.
Power: A New Social Analysis 1938 (New York: Norton)

9   With equal passion I have sought knowledge. I have wished to understand

the hearts of men. I have wished to know why the stars shine. And I have tried to apprehend the Pythagorean power by which number holds sway above the flux. A little of this, but not much, I have achieved.

The Autobiography of Bertrand Russell (London: Allen & Unwin) Introduction

Henry Norris Russell 1877–1957

1   The pursuit of an idea is as exciting as the pursuit of a whale.

[Lord] Ernest Rutherford 1871–1937

2   Don't let me catch anyone talking about the Universe in my department.

John Kendrew, BBC-3, 26 July 1968, 21.00 h
J D Bernal and the origin of life

3   If your experiment needs statistics, you ought to have done a better experiment.

in N T J Bailey *The Mathematical Approach to Biology and Medicine* 1967 (New York: Wiley)

4   It is essential for men of science to take an interest in the administration of their own affairs or else the professional civil servant will step in— and then the Lord help you.

Bulletin of the Institute of Physics 1950 **1** no. 1, cover

5   We haven't the money, so we've got to think.

in R V Jones *Bulletin of the Institute of Physics* 1962 **13** 102

6   [In answer to Stephen Leacock's enquiry as to what he thought of Einstein's theory of relativity] Oh, that stuff. We never bother with that in our work.

in Stephen Leacock *Common Sense and the Universe*

7   The energy produced by the breaking down of the atom is a very poor kind of thing. Anyone who expects a source of power from the transformation of these atoms is talking moonshine.

Physics Today 1970, October, p33

Gilbert Ryle 1900–

8   [There is no] Ghost in the Machine.

The Concept of Mind 1949 (London: Hutchinson)

George Alban Sacher 1917–

9   The brain is the organ of longevity.

[By its capacity to regulate the *milieu intérieur*]
Perspectives in Experimental Gerontology 1966 (Springfield, Ill: Thomas)

Antoine de Saint-Exupéry 1900–1944

1   *L'avion est une machine sans doute, mais quel instrument d'analyse! Cet instrument nous a fait découvrir le vrai visage de la terre* . . . . *Nous voilà donc changés en physiciens, en biologistes, examinant ces civilisations qui ornent des fonds de vallées* . . . . *Nous voilà donc jugeant l'homme a l'échelle cosmique, l'observant a travers nos hublots, comme à travers des instruments d'étude. Nous voilà relisant notre histoire.*
The aeroplane is, of course, only a machine, but what an instrument of analysis! This instrument has made us see the real face of the earth . . . .
Up here we are turned into physicists or biologists, studying the civilizations which garnish the depths of the valleys . . . . Up here we are judging man on a cosmic scale, observing him through our portholes, as through scientific instruments. We are re-reading our own history.
Oeuvres d'Antoine de Saint-Exupéry 1953 (Paris: Gallimard)

2   When the body sinks into death, the essence of man is revealed. Man is a knot, a web, a mesh into which relationships are tied. Only those relationships matter. The body is an old crock that nobody will miss. I have never known a man to think of himself when dying. Never.
Flight to Arras transl Lewis Galantiere (London: Heinemann)

Sakuma Shozan 1811–1864

3   *Toyo dotoku, seiyo geijutsu.*
Eastern ethics, Western techniques.
[The conditions for modernization at the Meiji Restoration]
[See Chang Chih-tung]

Abdus Salam 1926–

4   Al Asuli writing in Bukhara some 900 years ago divided his pharmacopoeia into two parts, 'Diseases of the rich' and 'Diseases of the poor'.
Scientific World 1963, no. 3, p9

Denis de Sallo 1626–1669

5   [First editor of the first scientific journal, writing in the first issue] *Personne ne doit trouver estrange de voir ici des opinions différentes des siennes, touchant les sciences, puisqu'on fait profession de rapporter les sentiments des autres sans les garantir* . . .
Nobody should find it strange to see here opinions different from his own concerning the sciences, because we aim to report the ideas of others without guaranteeing them . . .
Journal des Scavans 1665

Saif-ud-din Salman 15th Century

6   [Working at the observatory of Ulugh Beg in Samarkand] Admonish me not, my beloved father, for forsaking you thus in your old age and sojourning here at Samarkand. It is not that I covet the musk melons and the grapes and the pomegranates of Samarkand; it is not the shades of the orchards on the banks of Zar-Afsham, that keep me here. I love my native Kandahar and its tree-lined avenues even more and I pine to return. But

forgive me, my exalted father, for my passion for knowledge. In Kandahar there are no scholars, no libraries, no quadrants, no astrolabes. My star-gazing excites nothing but ridicule and scorn. My countrymen care more for the glitter of the sword than for the quill of the scholar. In my own town I am a sad, a pathetic misfit. It is true, my respected father, so far from home men do not rise from their seats to pay me homage when I ride into the bazaar. But some day soon, all Samarkand will rise in respect when your son will emulate Biruni and Tusi in learning and you too will feel proud.

[Abdus Salam tells us that, alas! Salman never did get his PhD]
transl Abdus Salam in *Minerva* 1966 **4** 461–5

Carl Sandburg 1878–1967

1   When water turns ice does it remember one time it was water?
When ice turns back into water does it remember it was ice?

Metamorphosis in *Honey and Salt* 1963
(New York: Harcourt, Brace & World)
© Carl Sandburg, 1963

George Santayana 1863–1952

2   The empiricist . . . thinks he believes only what he sees, but he is much better at believing than at seeing.
Skepticism and Animal Faith 1955 (New York: Dover)

3   Those who cannot remember the past are condemned to repeat it.
The Life of Reason 1905 (New York: Scribners)

George Alfred Leon Sarton 1884–1956

4   *Definition:* Science is systematised positive knowledge, or what has been taken as such at different ages and in different places.
Theorem: The acquisition and systematisation of positive knowledge are the only human activities which are truly cumulative and progressive.

Corollary: The history of science is the only history which can illustrate the progress of mankind. In fact, progress has no definite and unquestionable meaning in other fields than the field of science.
The Study of the History of Science 1957 (New York: Dover)

1   It is true that most men of letters and, I am sorry to add, not a few scientists, know science only by its material achievements, but ignore its spirit and see neither its internal beauty nor the beauty it extracts from the bosom of nature. Now I would say that to find in the works of science of the past, that which is not and cannot be superseded, is perhaps the most important part of our quest. A true humanist must know the life of science as he knows the life of art and the life of religion.
A History of Science vol II, 1959 (New York: Wiley)

2   Scientific activity is the only one which is obviously and undoubtedly cumulative and progressive.
The History of Science and the History of Civilization 1930 (New York: Dover)

3   The great intellectual division of mankind is not along geographical or racial lines, but between those who understand and practice the experimental method and those who do not understand and do not practice it.
in B Farrington *Science and Politics in the Ancient World* 1965 (London: Allen & Unwin)

George Savile [Lord Halifax] 1633–1695

4   He that leaveth nothing to chance will do few things ill, but he will do very few things.
Complete Works of George Savile ed W Raleigh, 1912

5   The struggle for knowledge hath a pleasure in it like that of wrestling with a fine woman.
Complete Works of George Savile ed W Raleigh, 1912

Friedrich von Schiller 1759–1805

6   *Einem ist sie [Wissenschaft] die hohe, die himmlische Göttin, dem anderen*
Eine tüchtige Kuh, die ihn mit Butter versorgt.
To one science is an exalted goddess; to another it is a cow which provides him with butter.
Xenien

7   *Nur die Fülle führt zur Klarheit,*
Und im Abgrund wohnt die Wahrheit.
Only wholeness leads to clarity,
And truth lies in the abyss.
[Favourite saying of Niels Bohr]

Erwin Schroedinger 1887–1961

8   . . . a living organism . . . feeds upon negative entropy . . . . Thus the device by which an organism maintains itself stationary at a fairly high level of

orderliness (= fairly low level of entropy) really consists in continually sucking orderliness from its environment.

What is life? 1944 (London: Cambridge UP)

Scottish Proverb

1   Guid gear gangs intae sma bouk.

Lucius Annaeus Seneca 4 BC–AD 65

2   In my own time there have been inventions of this sort, transparent windows, tubes for diffusing warmth equally through all parts of a building, short-hand, which has been carried to such a perfection that a writer can keep pace with the most rapid speaker. But the inventing of such things is drudgery for the lowest slaves; philosophy lies deeper. It is not her office to teach men how to use their hands. The object of her lessons is to form the soul. *Non est, inquam, instrumentorum ad usus necessarios opifex.*

Epistolae morales 90

3   [Vice] is still in its infancy, and yet on it we bestow all our efforts: our eyes and our hands are its slaves. Who attends the school of wisdom now? . . . Who has regard for philosophy or any liberal pursuit, except when a rainy day comes round to interrupt the games, and it may be wasted without loss? And so the many sects of philosophers are all dying out for lack of successors. The Academy, both old and new, has left no disciple.

in John Clarke *Physical Science in the Times of Nero: Being a Translation of the Quaestiones Naturales of Seneca* 1910 (London: Macmillan)

Severinus 7th Century

4   Go, my sons, buy stout shoes, climb the mountains, search . . . the deep recesses of the earth . . . . In this way and in no other will you arrive at a knowledge of the nature and properties of things.

William Shakespeare 1564–1616

5   . . . we have our philosophical persons, to make modern and familiar things supernatural and causeless.

All's Well that Ends Well II, iii

6   Plutus himself,
That knows the tinct and multiplying med'cine,
Hath not in nature's mystery more science
Than I have in this ring.

All's Well That Ends Well V, iii

7   It is as easy to count atomies, as to resolve the propositions of a lover.

As You Like It III, ii

8   . . . a traitorous innovator,
A foe to the public weal.

Coriolanus III, i

1   What doth gravity out of his bed at midnight?
    1 Henry IV II, iv

2   And time that takes survey of all the world
    Must have a stop.
    1 Henry IV V, iv

3   When we mean to build,
    We first survey the plot, then draw the model;
    And when we see the figure of the house,
    Then we must rate the cost of the erection; . . .
    2 Henry IV I, iii

4   Even so our houses and ourselves and children
    Have lost, or do not learn from want of time,
    The sciences that should become our country;
    But grow like savages, . . .
    Henry V V, ii

5   Thou hast most traitrously corrupted the youth
    Of the realm in erecting a grammar-school.
    2 Henry VI IV, vii

6   And nature must obey necessity.
    Julius Caesar IV, iii

7   If you can look into the seeds of time,
    And say which grain will grow and which will not,
    Speak then to me . . .
    Macbeth I, iii

8   . . . we but teach
    Bloody instructions, which being taught, return
    To plague the inventor.
    Macbeth I, vii

9   *Macbeth:* [The labour we delight] in physics pain.
    Macbeth II, iii

10   For there was never yet philosopher
     That could endure the toothache patiently.
     Much Ado about Nothing V, i

11   . . . physics the subject, makes old hearts fresh; . . .
     The Winter's Tale I, i

12   *Sir Toby:* Does not our lives consist of the four elements?
     Sir Andrew: Faith, so they say; but I think it rather consists of eating and

drinking.
Sir Toby: Th'art a scholar; let us therefore eat and drink.
Twelfth Night II, iii

George Bernard Shaw 1856–1950

1   All problems are finally scientific problems.
in Preface to *The Doctor's Dilemma*

2   The fact that a believer is happier than a sceptic is no more to the point than the fact that a drunken man is happier than a sober one. The happiness of credulity is a cheap and dangerous quality.
in Preface to *Androcles and the Lion*

3   Getting patronage is the whole art of life. A man cannot have a career without it.
Captain Brassbound's Conversion 1906, Act III

4   Great art is never produced for its own sake. It is too difficult to be worth the effort.
in Preface to *Three Plays by Brieux*

5   Tyndall declared that he saw in Matter the promise and potency of all forms of life, and with his Irish graphic lucidity made a picture of a world of magnetic atoms, each atom with a positive and a negative pole, arranging itself by attraction and repulsion in orderly crystalline structure. Such a picture is dangerously fascinating to thinkers oppressed by the bloody disorders of the living world. Craving for purer subjects of thought, they find in the contemplation of crystals and magnets a happiness more dramatic and less childish than the happiness found by mathematicians in abstract numbers, because they see in the crystals beauty and movement without the corrupting appetites of fleshly vitality.
in Preface to *Back to Methuselah*

[Bishop] Fulton Sheen 1895–

6   An atheist is a man who has no invisible means of support.
Look 14 December 1955

Charles S Sheldon

7   What is the long-run psychological cost to us of having the backside of the Moon dotted with Soviet names? Will they do the same for Mars? To pretend that national prestige is unimportant is to show a limited awareness of historical forces in society.
National Goals in Space, NASA in *Proceedings of the Working Conference on Space Nutrition and Related Waste Problems, Tampa, Florida*

Percy Bysshe Shelley 1792–1822

8   . . . happiness

And science dawn though late upon the earth; . . .
Queen Mab VIII, 11.227–8

1   The gigantic shadows which futurity casts upon the present.
The Defence of Poetry

2   He gave man speech, and speech created thought,
Which is the measure of the universe;
And Science struck the thrones of earth and heaven.
Prometheus Unbound 1820

Herbert Alexander Simon 1916–

3   A man, viewed as a behaving system, is quite simple. The apparent complexity of his behaviour over time is largely a reflection of the complexity of the environment in which he finds himself.
The Sciences of the Artificial 1969 (Cambridge, Mass: The MIT Press)

[Sir] John Sinclair 1754–1835

4   At present there are a greater number of intelligent practical chemists in Scotland, in proportion to the population, than perhaps in any other country in the world.
1814

Burrhus Frederic Skinner 1904–

5   'The science of behaviour is full of special twists like that,' said Frazier. 'It's the science of science a special discipline concerned with talking about talking and knowing about knowing. Well, there's a motivational twist too. Science in general emerged from a competitive culture. Most scientists are still inspired by competition or at least supported by those who are. But when you come to apply the methods of science to the special study of human behaviour, the competitive spirit commits suicide. It discovers the extraordinary fact that in order to survive, we must in the last analysis, not compete.'
Walden Two 1948 (New York: Macmillan)

6   Education is what survives when what has been learnt has been forgotten.
New Scientist 21 May 1964

Martyn Skinner 1906–

7   For we are like the Chinese in reverse.
Our feeling for the future's so prodigious
We might be termed descendant worshippers.
The Return of Arthur 1966 (London: Chapman & Hall)

Samuel Smiles 1812–1904

8   We often discover what will do, by finding out what will not do; and probably he who never made a mistake never made a discovery.
Self-help ch 11

Adam Smith 1723–1790

1   The machines that are first invented to perform any particular movement are always the most complex, and succeeding artists generally discover that with fewer wheels, with fewer principles of motion than had originally been employed, the same effects may be more easily produced. The first philosophical systems, in the same manner, are always the most complex.
Essay on the Principles which Lead and Direct Philosophical Inquiries

2   Science is the great antidote to the poison of enthusiasm and superstition.
Wealth of Nations V, part 3.3

Cyril Stanley Smith 1903–

3   Matter is a holograph of itself in its own internal radiation.
Letter to A L Mackay, 1968

Sydney Smith 1771–1845

4   Science is his forte, and omniscience his foible.
[Of William Whewell]
in Isaac Todhunter *William Whewell* (Farnborough: Gregg International)

[General] Walter Bedell Smith 1895–1961

5   My big job is to get the best brains in the country, persuade them to leave fame and fortune for a government job where they'll study secrets they can't even discuss with their wives . . .
[Second Director of the CIA]
They call it intelligence 1963 (London: Abelard–Schuman)

[Lord] Charles Percy Snow 1905–

6   A good many times I have been present at gatherings of people who, by the standards of traditional culture, are thought highly educated and who have with considerable gusto been expressing their incredulity at the illiteracy of scientists. Once or twice I have been provoked and have asked the company how many of them could describe the Second Law of Thermodynamics. The response was cold: it was also negative.
The Two Cultures The Rede Lecture, 1959 (London: Cambridge UP)

7   The scientific revolution is the only method by which most people can gain the primal things (years of life, freedom from hunger, survival for children) —the primal things which we take for granted and which have in reality come to us through having had our own scientific revolution not so long ago.
The Two Cultures: A Second Look 1963 (London: Cambridge UP)

8   Scientists have it within them to know what a future directed society feels like, for science itself, in its human aspect, is just like that.
A Postscript to Science and Government 1962 (Oxford: Oxford UP)

9   Scientists . . . . I should say that naturally they had the future in their

bones.
The Two Cultures The Rede Lecture, 1959 (London: Cambridge UP)

1   [Of molecular biology] This branch of science is likely to affect the way in which men think of themselves more profoundly than any scientific advance since Darwin's and probably more so than Darwin's.
The Two Cultures: A Second Look 1963 (London: Cambridge UP)

Frederick Soddy 1877–1956

2   Four circles to the kissing come,
       The smaller are the benter.
       The bend is just the inverse of
       The distance from the centre.
       Though their intrigue left Euclid dumb
       There's now no need for rule of thumb.
       Since zero bend's a dead straight line
       And concave bends have minus sign,
       The sum of squares of all four bends
       Is half the square of their sum.
       Nature 1936 **137** 1021

Omond McKillop Solandt 1909–

3   It is presumptuous for scientists to try to formulate national goals, since science is by no means the only important activity in the nation. But scientists have a duty to point out that most nations have neither explicit goals nor a mechanism for formulating them.

Sophocles 495–406 BC

4   One must learn by doing the thing; though you think you know it, you have no certainty until you try.
       Trachiniae 592

Herbert Spencer 1820–1903

5   Definition of life: The continuous adjustment of internal relations to external relations.
       Principles of Biology section 30

6   Science is organised knowledge.
       Education ch 29

Stephen Spender 1909–

7   More beautiful and soft than any moth
       With burring furred antennae feeling its huge path
       Through dusk, the airliner with shut-off engines
       Glides over suburbs and the sleeves set trailing tall
       To point the wind. Gently, broadly she falls,
       Scarcely disturbing charted currents of air.
       The Landscape near an Aerodrome in *Collected Poems, 1928–53* (London: Faber & Faber)

Oswald Spengler 1880–1936

1   Nature is the shape in which the man of higher cultures synthesizes and interprets the immediate impressions of his senses. History is that from which his imagination seeks comprehension of the living existence of the world.
The Decline of the West transl C F Atkinson, vol 1, 1926 (London: Allen & Unwin)

Edmund Spenser *ca* 1552–1599

2   Within this wide great Universe
Nothing doth firme and permanent appeare,
But all things tost and turned by transverse.
Two Cantos of Mutability 1609

State of Tennessee, USA

3   It shall be unlawful for any teacher in any of the universities, normals and all other public schools of the state which are supported in whole or in part by the public school funds of the state, to teach any theory that denies the story of the divine creation of man as taught in the Bible, and to teach instead that man has descended from a lower order of animals.
1925. Repealed 1967

Gertrude Stein 1874–1946

4   When we were having a book printed in France we complained about the bad alignment. Ah, they explained, that is because they use machines now, machines are bound to be inaccurate, they have not the intelligence of human beings, naturally the human mind corrects the fault of the hand, but with a machine, of course, there are errors. The reason why all of us naturally began to live in France is because France has scientific methods, machines and electricity, but does not really believe that these things have anything to do with the real business of living.
Paris France 1940 (New York: Scribners)

Stendhal [Henri Beyle] 1783–1842

5   *Ce que j'appelle cristallisation, c'est opération de l'esprit, qui tire de tout ce qui se présente la découverte que l'objet aimé a des nouvelles perfections.*
I call 'crystallisation' that action of the mind that discovers fresh perfections in its beloved at every turn of events.
De l'amour 1822, ch 1

Gunther Siegmund Stent 1924–

6   It would, of course, be a poor lookout for the advancement of science if young men started believing what their elders tell them, but perhaps it is legitimate to remark that young Turks look younger, or more Turkish, . . . if the conclusions they eventually reach are different from what anyone had said before.
Nature 1969 **221** 320

George Stephenson 1781–1848

1   I will send them the locomotive to be the Great Missionary among them.

Laurence Sterne 1713–1768

2   It is in the nature of a hypothesis when once a man has conceived it, that it assimilates everything to itself, as proper nourishment, and from the first moment of your begetting it, it generally grows stronger by everything you see, hear or understand.
Tristram Shandy 1759–1767

3   Through prolonged close contact and friction with the objects of their study, the minds of experts finally acquire a pictorial, moth-like, fiddling perfection.

Wallace Stevens 1879–1955

4   The man bent over his guitar,
     A shearsman of sorts. The day was green.

They said, 'You have a blue guitar,
You do not play things as they are'.
The man replied, 'Things as they are
Are changed upon a blue guitar . . .
The Man with the Blue Guitar in *Collected Poems of Wallace Stevens* (London: Faber & Faber)

Adlai E Stevenson 1900–1965

1   *Via ovicipitum dura est.*
    The way of the egghead is hard.
    Attributed

Strabo 1st Century BC

2   The poets were not alone in sponsoring myths. Long before them cities and
    lawmakers had found them a useful expedient . . . . They needed to control
    the people by superstitious fears, and these cannot be aroused without
    myths and marvels.
    Geography I, 2; 8

Alexander Strange 1818–1876

3   1. That science is essential to the advancement of civilisation, the develop-
    ment of national wealth, and the maintenance of national power.
    2. That all science should be cultivated, even branches of science which do
    not appear to promise immediately direct advantage.
    3. That the State or Government, acting as trustees of the people, should
    provide for the cultivation of those departments of science which, by reason
    of costliness, either in time or money, or of remoteness of probable profit,
    are beyond the reach of private individuals; in order that the community
    may not suffer from the effect of insufficiency of isolated effort.
    4. That to whatever extent science may be advanced by State agency, that
    agency should be systematically constituted and directed.
    Conclusions of *Devonshire Royal Commission on Scientific Institutions and the Advancement o
    Science* 1872

Igor Stravinsky 1882–1973

4   The more constraints one imposes, the more one frees one's self of the
    chains that shackle the spirit . . . the arbitrariness of the constraint only
    serves to obtain precision of execution.
    Poetics of Music (Cambridge, Mass: Harvard UP)

Johann August Strindberg 1849–1912

5   Then is it reasonable to think that one can see, by looking in a microscope,
    what is going on in another planet?
    The Father 1887, Act 1, scene 5

Sun Tze 5th–6th Century BC

6   Hence to fight and conquer in all your battles is not supreme excellence;
    supreme excellence consists in breaking the enemy's resistance without
    fighting.
    [Favourite author of Mao Tse-tung and many others]
    Sun Tze Ping Fa transl L Giles, 1910 (London: Luzac)

Jonathan Swift 1667–1745

1   If they would, for example, praise the
     beauty of a woman, or any other animal,
     they describe it by rhombs, circles,
     parallelograms, ellipses, and other
     geometrical terms . . .
     A Voyage to Laputa in *Gulliver's Travels*

2   In the school of political projectors, I was
     but ill entertained, the professors appearing,
     in my judgment, wholly out of their senses;
     which is a scene that never fails to make
     me melancholy. These unhappy people
     were proposing schemes for persuading
     monarchs to choose favourites upon the
     score of their wisdom, capacity, and virtue;
     of teaching ministers to consult the public
     good; of rewarding merit, great abilities,
     and eminent services; of instructing
     princes to know their true interest, by
     placing it on the same foundation with
     that of their people; of choosing for
     employment persons qualified to exercise
     them; with many other wild impossible
     chimeras, that never entered before into
     the heart of man to conceive, and confirmed in me the old observation,
     that there is nothing so extravagant and irrational which some philosophers
     have not maintained for truth.
     A Voyage to Laputa in *Gulliver's Travels*

3   . . . whoever could make two ears of corn, or two blades of grass, to grow
     upon a spot of ground where only one grew before, would deserve better of
     mankind, and do more essential service to his country, than the whole race
     of politicians put together.
     A Voyage to Brobdingnag in *Gulliver's Travels*

4   [Of the Laputans] They have likewise discovered two lesser stars, or satel-
     lites, which revolve about Mars, whereof the innermost is distant from
     the centre of the primary planet exactly three of his diameters, and the
     outermost five; the former revolves in the space of ten hours, and the latter
     in twenty one and a half; . . .
     [These satellites were first observed by Asaph Hall in 1877. Their periods are 7·7 and 30 hours]
     A Voyage to Laputa in *Gulliver's Travels*

Albert Szent-Gyorgyi 1893–

5   Discovery consists of seeing what everybody has seen and thinking what
     nobody has thought.
     The Scientist Speculates ed I G Good, 1962 (London: Heinemann)

6   It is common knowledge that the ultimate source of all our energy and

negative entropy is the radiation of the sun. When a photon interacts with a material particle on our globe it lifts one electron from an electron pair to a higher level. This excited state as a rule has but a short lifetime and the electron drops back within 10^{-7} to 10^{-8} seconds to the ground state giving off its excess energy in one way or another. Life has learned to catch the electron in the excited state, uncouple it from its partner and let it drop back to the ground state through its biological machinery utilizing its excess energy for life processes.

Light and Life ed W D McElroy and B Glass, 1961 (Baltimore, Md: Johns Hopkins Press)

1   Knowledge is a sacred cow, and my problem will be how we can milk her while keeping clear of her horns.
Science 1964 **146** 1278

2   The real scientist . . . is ready to bear privations and, if need be, starvation rather than let anyone dictate to him which direction his work must take.
Science Needs Freedom 1943

3   Research means going out into the unknown with the hope of finding something new to bring home. If you know in advance what you are going to do, or even to find there, then it is not research at all: then it is only a kind of honourable occupation.
Perspectives in Biology and Medicine 1971, p1

Leo Szilard 1898–1964

4   Don't lie if you don't have to.
Science 1972 **176** 966

Taibai died 1842

5   To a man who has not eaten a globe-fish, we cannot speak of its flavour.
[This is the *fugu*, parts of which are extremely poisonous]

Hippolyte Taine 1828–1893

6   *Le vice et la vertu sont des produits comme le vitriol et le sucre.*
Vice and virtue are products like sulphuric acid and sugar.
Histoire de la littérature anglaise 1863, Introduction

Peter Guthrie Tait 1831–1901

7   Perhaps to the student there is no part of elementary mathematics so repulsive as is spherical trigonometry.
Encyclopaedia Britannica 1911, 11th edn, 22, 721 (article on *Quaternions*)

Charles-Maurice De Talleyrand 1754–1838

8   Both erudition and agriculture ought to be encouraged by government; wit and manufactures will come of themselves.

9   It is sometimes quite enough for a man to feign ignorance of that which he knows, to gain the reputation of knowing that of which he is ignorant.

Vasili N Tatishchev 1686–1750

1   Freedom is not an essential and basic condition for the growth of science; the care and diligence of government authorities are the most important conditions for this development.
Razgovor

Alfred [Lord] Tennyson 1809–1892

2   . . . . Saw the heavens fill with commerce, argosies of magic sails,
Pilots of the purple twilight, dropping down with costly bales;
Heard the heavens fill with shouting, and there rained a ghastly dew
From the nations' airy navies grappling in the central blue . . . .
Science moves, but slowly slowly, creeping on from point to point
Slowly comes a hungry people, as a lion creeping nigher,
Glares at one that nods and winks behind a slowly-dying fire.
. . . . Better fifty years of Europe than a cycle of Cathay.
[Prophetic—which excuses the ghastly rhyme]
Locksley Hall 1832

3   Science grows and Beauty dwindles.
Locksley Hall Sixty Years After 1886

4   A time to sicken and to swoon,
When Science reaches forth her arms
To feel from world to world, and charms
Her secret from the latest moon.
In Memoriam AHH XXI, 4

Tertullian *ca* 155–222

5   *Certum est quia imposibile est.*
It is certain because it is impossible.
De Carne Cristi 5

Thales of Miletus *ca* 640–546 BC

6   Water is the principle, or the element, of things
All things are water.
Plutarch *Placita Philosophorum* i, 3

Stefan Themerson 1910–

7   With a poet it's different,
if his poem is bad
even his broken heart
will not make it better.
On Semantic Poetry 1975 (London: Gaberbocchus)

D'Arcy Wentworth Thompson 1860–1948

8   Cell and tissue, shell and bone, leaf and flower, are so many portions of matter, and it is in obedience to the laws of physics that their particles have been moved, moulded and conformed. They are no exception to the rule that God always geometrizes. Their problems of form are in the first

instance mathematical problems, their problems of growth are essentially physical problems, and the morphologist is, *ipso facto*, a student of physical science.
On Growth and Form 1917 (London: Cambridge UP)

1   Form is a diagram of forces.
On Growth and Form 1917 (London: Cambridge UP)

2   It behoves us always to remember that in physics it has taken great men to discover simple things. They are very great names indeed which we couple with the explanation of the path of a stone, the droop of a chain, the tints of a bubble, the shadows in a cup.
On Growth and Form 1917 (London: Cambridge UP)

Francis Thompson 1859–1907

3   All things by immortal power,
Near and far
Hiddenly
To each other linked are,
That thou canst not stir a flower
Without troubling of a star.
The Mistress of Vision

James Thomson 1700–1748

4   Even now the setting sun and shifting clouds,
Seen, Greenwich, from thy lovely heights, declare
How just, how beauteous the refractive law.
To the Memory of Sir Isaac Newton 1727

5   [Newton] from motion's simple laws
Could trace the secret hand of Providence,
Wide-working through this universal frame.
To the Memory of Sir Isaac Newton 1727

[Sir] Joseph John Thomson 1856–1940

6   This example illustrates the differences in the effects which may be produced by research in pure or applied science. A research on the lines of applied science would doubtless have led to improvement and development of the older methods—the research in pure science has given us an entirely new and much more powerful method. In fact, research in applied science leads to reforms, research in pure science leads to revolutions, and revolutions, whether political or industrial, are exceedingly profitable things if you are on the winning side.
in Lord Rayleigh J J Thomson 1943 (London: Cambridge UP)

William Thomson [Lord Kelvin] 1824–1907

7   Do not imagine that mathematics is hard and crabbed, and repulsive to common sense. It is merely the etherialization of common sense.
in S P Thompson Life of Lord Kelvin 1910

1   Fourier is a mathematical poem.
W Thompson and P G Tait *Treatise on Natural Philosophy* vol I, pp713, 718. Quoted by F Engels in *The Dialectics of Nature*

2   I am never content until I have constructed a mechanical model of the subject I am studying. If I succeed in making one, I understand; otherwise I do not.
Notes of Lectures on *Molecular Dynamics and the Wave Theory of Light*

3   I often say that when you can measure what you are speaking about, and express it in numbers, you know something about it; but when you cannot measure it, when you cannot express it in numbers, your knowledge is of a meagre and unsatisfactory kind.
Lecture to the Institution of Civil Engineers, 3 May 1883

Henry David Thoreau 1817–1862

4   It appears to be a law that you cannot have a deep sympathy with both man and nature.
Walden

5   Simplicity, simplicity, simplicity. I say, let your affairs be as two or three, and not a hundred or a thousand; instead of a million count half a dozen, and keep your accounts on your thumb nail.
Walden. See *Scientific American* August 1969

James Thurber 1894–1961

6   Progress was all right; it only went on too long.
Attributed

Kliment Arkadievich Timiryazev 1843–1920

7   I set myself two parallel tasks: to create for science and write for the people.
Science and Democracy 1927 (Leningrad: Priboi)

[Sir] Henry Tizard 1885–1959

8   Andrade [who was looking after wartime inventions] is like an inverted Micawber, waiting for something to turn down.
in C P Snow *A Postscript to Science and Government* 1962 (Oxford: Oxford UP)

9   The secret of science is to ask the right question, and it is the choice of problem more than anything else that marks the man of genius in the scientific world.
in C P Snow *A Postscript to Science and Government* 1962 (Oxford: Oxford UP)

[Count] Lev Nikolaevich Tolstoi 1828–1910

10   I am convinced that the history of so-called scientific work in our famous centuries of European civilisation will, in a couple of hundred years, represent an inexhaustible source of laughter and sorrow for future generations. The learned men of the small western part of our European continent

lived for several centuries under the illusion that the eternal blessed life was the West's future. They were interested in the problem of when and where this blessed life would come. But they never thought of how they were going to make their life better.

1884. Probably in *What is Religion?*

1 Our body is a machine for living.

Napoleon in *War and Peace* transl L and A Maude, Book X, 1922 (Oxford: Oxford UP) ch 29

2 A modern branch of mathematics, having achieved the art of dealing with the infinitely small, can now yield solutions in other more complex problems of motion, which used to appear insoluble.

This modern branch of mathematics, unknown to the ancients, when dealing with problems of motion, admits the conception of the infinitely small, and so conforms to the chief condition of motion (absolute continuity) and thereby corrects the inevitable error which the human mind cannot avoid when dealing with separate elements of motion instead of examining continuous motion.

In seeking the laws of historical movement just the same thing happens. The movement of humanity, arising as it does from innumerable human wills, is continuous.

To understand the laws of this continuous movement is the aim of history . . . .

Only by taking an infinitesimally small unit for observation (the differential of history, that is, the individual tendencies of men) and attaining to the art of integrating them (that is, finding the sum of these infinitesimals) can we hope to arrive at the laws of history.

War and Peace transl L and A Maude, Book XI, 1922 (Oxford: Oxford UP) ch 1

3 The generals, the institution can select a strategy, lay it all out, but what happens on the battlefield is quite different.

4 [Of science] It gives us no answer to our question, what shall we do and how shall we live?

What is Art? 1898

Rudolf Tomaschek 1895–

5 Modern physics is an instrument of Jewry for the destruction of Nordic science . . . . True physics is the creation of the German spirit.

in W L Shirer *The Rise and Fall of the Third Reich* (London: Secker & Warburg) ch 8

Stephen Toulmin 1922–

6 No doubt, a scientist isn't necessarily penalized for being a complex, versatile, eccentric individual with lots of extra-scientific interests. But it certainly doesn't help him a bit.

Civilization and Science in Conflict or Collaboration 1972 (Amsterdam: Elsevier)

Arnold Toynbee 1889–1975

7 [We are in] the first age since the dawn of civilisation in which people

have dared to think it practicable to make the benefits of civilisation
available to the whole human race.

Thomas Traherne *ca* 1637–1674

1   He that knows the secrets of nature with Albertus Magnus, or the motions
of the heavens with Galileo, or the cosmography of the moon with
Hevelius, or the body of man with Galen, or the nature of diseases with
Hippocrates, or the harmonies in melody with Orpheus, or of poetry with
Homer, or of grammar with Lily, or of whatever else with the greatest
artist; he is nothing if he knows them merely for talk or idle speculation,
or transient and external use. But he that knows them for value, and knows
them his own, shall profit infinitely.
Centuries of Meditation 1908, no. 341

Lev Davidovich [Bronstein] Trotskii 1879–1940

2   The phenomena of radio-activity lead us straight to the problem of
releasing the inner energy of the atom . . . . The greatest task of contemporary physics is to extract from the atom its latent energy—to tear open a
plug so that energy should well up with all its might. Then it will become
possible to replace coal and petrol by atomic energy which will become our
basic fuel and motive power. This is by no means a hopeless task, and what
vistas its solution will open up . . . scientific and technological thought is
approaching the point of a great upheaval; and so the social revolution of
our time coincides with a revolution in man's inquiry into the nature of
matter and in his mastery of matter.
Speech 1 March 1926. *Sochineniya* XXI

3   From the field of chemistry there is no direct and immediate exit to social
perspectives . . . . An objective method of social cognition is necessary.
Marxism is that method. When any marxist tried to convert Marx's theory
into a universal skeleton key and flitted through other fields of knowledge,
Vladimir Il'ich would rebuke him with the expressive little phrase,
'Communist conceit'. This would signify in particular: Communism does
not replace chemistry. But the converse is also true. The attempt to step
across Marxism, on the pretext that chemistry (or natural science in
general) must solve all problems, is a peculiar chemical conceit, which is
theoretically no less erroneous and practically no more likeable than
Communist conceit.
D I Mendeleev i Marksizm in *Sochineniya* XXI

Harry S Truman 1884–1972

4   *Oppenheimer allant dans le bureau de Truman avec Dean Acheson disait à
ce dernier en se tordant les mains 'J'ai du sang sur les mains' et que plutard
Truman dit à Acheson: 'ne me ramenez plus jamais ce f. . . cretin. C' n'est
pas lui qui a lancé la bombe. C'est moi. Cette sorte de pleurnicherie me rend
malade.'*
[The initial 'S' in Harry S Truman does not stand for anything]
in Jean-Jacques Salomon *Science et Politique* 1970 (Paris: Editions du Seuil)

Ivan Sergeievich Turgenev 1818–1883

1   Nature is not a temple but a workshop in which man is the labourer.

2   Whatever a man prays for, he prays for a miracle. Every prayer reduces itself to this: 'Great God, grant that twice two be not four'.
Prayer

Turkish Proverb

3   If Allah gives you prosperity, He will give you the brains to go with it.

Mark Twain [Samuel Langhorne Clemens] 1835–1910

4   What a good thing Adam had—when he said a thing he knew nobody had said it before.

5   When I was a boy of 14 my father was so ignorant I could hardly stand to have the old man around. But when I got to be 21, I was astonished at how much he had learnt in 7 years.

6   Scientists have odious manners, except when you prop up their theory; then you can borrow money of them.
The Bee in *What is Man and Other Essays*

Henry Twells 1823–1900

7   When as a child I laughed and wept,
        Time crept.
    When as a youth I waxed more bold,
        Time *strolled.*
    When I became a full-grown man,
        Time RAN.
    When older still I daily grew,
        Time FLEW.
    Soon I shall find, in passing on,
        Time *gone.*
    O Christ! wilt Thou haved saved me then?
        Amen.
[Poem fixed to the front of the clock-case in the North Transept of Chester Cathedral]
Time's Paces in *Newsletter of the Friends of Chester Cathedral* Christmas 1972

United States Air Force

8   [The United States Air Force ROTC Manual *Fundamentals of Aerospace Weapons Systems* defines a MILITARY TARGET as] Any person, thing, idea, entity or location selected for destruction, inactivation, or rendering non-usable with weapons which will reduce or destroy the will or ability of the enemy to resist.
[Rapoport draws attention to the mentality which attacks ideas with bombs]
in K Von Clausewitz *On War* ed A Rapoport, 1967 (London: Routledge & Kegan Paul)

US President's Science Advisory Committee

1   In science the excellent is not just better than the ordinary; it is almost all that matters. It is therefore fundamental that this country should energetically sustain and strongly reinforce first-rate work where it now exists.
Scientific Progress, the Universities and the Federal Government The White House, Washington, DC, 15 November 1960

Miguel de Unamuno 1864–1937

2   Science is a cemetery of dead ideas.
The Tragic Sense of Life transl P Smith, 1953 (London: Routledge & Kegan Paul)

John Updike 1932–

3   Neutrinos, they are very small.
They have no charge and have no mass
And do not interact at all.
The earth is just a silly ball
To them, through which they simply pass,
Like dustmaids down a drafty hall
Or photons through a sheet of glass.
They snub the most exquisite gas,
Ignore the most substantial wall,
Cold shoulder steel and sounding brass,
Insult the stallion in his stall,
And, scorning barriers of class,
Infiltrate you and me. Like tall
And painless guillotines, they fall
Down through our heads into the grass.
At night, they enter at Nepal
And pierce the lover and his lass
From underneath the bed—you call
It wonderful; I call it crass.
Cosmic Gall in *Telegraph Poles and Other Poems* (London: Deutsch)

James Ussher [Archbishop of Armagh] 1581–1656

4   The world was created on 22nd October, 4004 BC at 6 o'clock in the evening.
Chronologia Sacra Oxford, 1660, p45. *Annals of the World* Oxford, 1658

Paul Valéry 1871–1945

5   'Science' means simply the aggregate of the recipes that are always successful. All the rest is literature.
Analects vol 14 of *Collected Works* ed J Matthews, 1970 (London: Routledge & Kegan Paul)

6   Having precise ideas often leads to a man doing nothing.

7   If a man's imagination is stimulated by artificial and arbitrary rules, he is a poet; if it is stifled by such limitations, whatever kind of writer he may be, a poet he is not.

1   *L'histoire est la science des choses qui ne se repetent pas.*
History is the science of things which are not repeated.
Variété IV

2   Man is only man at the surface. Remove his skin, dissect, and immediately you come to the machinery.

3   One had to be a Newton to notice that the moon is falling, when everyone sees that it doesn't fall.
Analects vol 14 of *Collected Works* ed J Matthews, 1970 (London: Routledge & Kegan Paul)

4   There is a *science* of simple things, an *art* of complicated ones. Science is feasible when the variables are few and can be enumerated; when their combinations are distinct and clear. We are tending toward the condition of science and aspiring to do it. The artist works out his own formulas; the interest of science lies in the *art* of making science.
Analects vol 14 of *Collected Works* ed J Matthews, 1970 (London: Routledge & Kegan Paul)

Giorgio Vasari 1511–1574

5   He might have been a scientist if he had not been so versatile.
[Of Leonardo da Vinci]
Lives of the Artists

The Vatican Council

6   If any one shall not be ashamed to assert that, except for matter, nothing exists; let him be anathema.
Session 3, Canon 2, 24 April 1870

Nikolai Ivanovich Vavilov 1887–1943

7   We shall go to the pyre, we shall burn, but we shall not renounce our convictions.
[The geneticist arrested 6 August 1940; sentenced to death 9 July 1941; elected Foreign Member of the Royal Society 1942; died 26 January 1943]
in Zh A Medvedev *The Rise and Fall of T D Lysenko* 1969 (New York: Columbia UP)

Sergei Ivanovich Vavilov 1891–1951

8   How have the thematics of scientific research at different times and places been determined and how are they determined? It is only today that we have begun to study this most important problem of the history of science and it is only the Marxists who are doing it.
Marxism and Modern Thought 1935 (London: Routledge)

Thorstein Veblen 1857–1929

9   The outcome of any serious research can only be to make two questions grow where only one grew before.
The Place of Science in Modern Civilization and Other Essays 1919 (New York: Viking Press)

Paul Verlaine 1844–1896

10   Brothers, touch not gluttonous science

That off the forbidden vines seeks to steal
The bloody fruit we must not know.
in Jacques Monod *Inaugural Lecture* 1967 (San Diego, Calif: The Salk Institute)

Jules Verne 1828–1905

1   As for the Yankees, they have no other ambition than to take possession
of this new continent of the sky, and to plant upon the summit of its highest elevation the star-spangled banner of the United States.

Madame Marie Vichy-Deffand [Marquise du Deffand] 1697–1780

2   *Il n'y a que le premier pas qui coute.*
It is only the first step which takes the effort.
[Referring to the legend of Saint Denis who walked from his place of execution carrying his head]
Lettre à d'Alembert 1763

[Sir] Geoffrey Vickers 1894–

3   . . . the historical causes which produced the Western individual and turned
him into the Western individualist. I will not elaborate them here. I would
only insist that we should not mistake for laws of God or nature the cultural values of the world's most unstable systems.
Freedom in a Rocking Boat 1970 (London: Penguin)

Rudolf Virchow 1821–1902

4   In my journal, anyone can make a fool of himself.
Archiv für pathologische Anatomie und Physiologie (Zeitschrift für Ethnologie)

5   Pathology is the science of disease [in all organisms] from cells to societies.
Archiv für pathologische Anatomie und Physiologie Introduction

Virgil [Virgilius Maro] 70–19 BC

6   *Exudent alii spirantia mollius aera*
Credo equidem vivos ducent de marmore vultus
Orabunt causas melius, caelique meatus
Describent radio, et surgentia sidera dicent.
Tu regere imperio populos, Romane, memento,
Hae tibi erunt artes, pacisque imponere morem
Parcere subjectis et debellare superbos.
Others, I know it well, breathing bronze shall trace
And from the deathlike marble call up the living face:
Shall plead with eloquence not thine, shall map and rule the skies,
And with the voice of science shall tell when stars shall set and rise.
'Tis thine O Rome to rule: this mission ne'er forgo
Thine arts, thy science this, to dictate to the foe
To spare who yields submission and bring the haughty low.
Aeneid VI, 85

7   *Felix qui potuit rerum cognoscere causas.*
Happy is he who gets to know the reasons for things.
[Motto of Churchill College, Cambridge—local translation: It's great to know what makes things tick]
Georgics II, 490

Voltaire [François Marie Arouet] 1694–1778

1   *Vous avez confirmé dans ces lieux pleins d'ennui*
Ce que Newton connut sans sortir de chez lui.
You have confirmed in these tedious places
What Newton found out without leaving his room.
[On the expedition of Maupertuis to Peru to confirm the flattening of the Earth at the poles]

2   [Invited a second time to an orgy] 'Ah no, my good friends, once a
philosopher, twice a pervert'.

3   There is an astonishing imagination, even in the science of mathematics
. . . . We repeat, there was far more imagination in the head of Archimedes
than in that of Homer.
A Philosophical Dictionary 1881, 3, 40

4   *'Travaillons sans raisonner,' dit Martin; 'c'est le seul moyen de rendre la vie*
supportable.'
'Let us work without theorising,' said Martin; ''tis the only way to make
life endurable.'
Candide 1758, XXX. Transl John Butt, 1947 (London: Penguin Classics)

5   *Cunégonde . . . vit entre les broussailles le docteur-Pangloss qui donnait une*
leçon de physique expérimentale à la femme de chambre de sa mère, petite
brune tres jolie et tres docile.
One day Cunégonde was walking near the house in a little coppice, called
'the park', when she saw Dr Pangloss behind some bushes giving a lesson
in experimental philosophy to her mother's waiting woman, a pretty little
brunette who seemed eminently teachable.
Candide 1758, I. Transl John Butt, 1947 (London: Penguin Classics)

[Field-Marshal] Helmuth Carl Bernard Von Moltke 1800–1891

6   No plan survives contact with the enemy.

Alexander Vucinich 1914–

7   Every scientist is an agent of cultural change. He may not be a champion
of change; he may even resist it, as scholars of the past resisted the new
truths of historical geology, biological evolution, unitary chemistry, and
non-Euclidean geometry. But to the extent that he is a true professional,
the scientist is inescapably an agent of change. His tools are the instruments
of change—skepticism, the challenge to established authority, criticism,
rationality, and individuality.
Science in Russian Culture: A History to 1860 1963 (Stanford, Calif: Stanford UP)

Conrad Hal Waddington 1905–1975

8   Science is the organised attempt of mankind to discover how things work
as causal systems. The scientific attitude of mind is an interest in such
questions. It can be contrasted with other attitudes, which have different

interests; for instance the magical, which attempts to make things work not as material systems but as immaterial forces which can be controlled by spells; or the religious, which is interested in the world as revealing the nature of God.

The Scientific Attitude 1941 (London: Penguin)

George Wald 1906–

1   We are the products of editing, rather than of authorship.

in *The Origin of Optical Activity* from the *Annals of the New York Academy of Sciences* 1957 **69** 352–68

Alfred Russell Wallace 1823–1913

2   In proportion as physical characteristics become of less importance, mental and moral qualities will have an increasing importance to the well-being of the race. Capacity for acting in concert, for protection of food and shelter; sympathy, which leads all in turn to assist each other; the sense of right, which checks depredation upon our fellows . . . all qualities that from earliest appearance must have been for the benefit of each community, and would therefore have become objects of natural selection.

Origin of human races and the antiquity of man in *Journal of the Royal Anthropological Society, London* 1864, clviii

3   These checks—war, disease, famine and the like—must, it occurred to me, act on animals as well as man. Then I thought of the enormously rapid multiplication of animals, causing these checks to be much more effective in them than in the case of man; and while pondering vaguely on this fact there suddenly flashed upon me the idea of the survival of the fittest—that the individuals removed by these checks must be on the whole inferior to those that survived. In the two hours that elapsed before my ague fit was over, I had thought out almost the whole of the theory: and the same evening I sketched the draft of my paper, and in the two succeeding evenings wrote it out in full, and sent it by the next post to Mr. Darwin.

in B Willey *Darwin and Butler* 1960 (London: Chatto & Windus)

Graham Wallas 1858–1932

4   . . . 'How can I know what I think till I see what I say?'

The Art of Thought ed May Wallas, 1945 (London: Watts)

Robert Penn Warren 1905–

5   . . . . What if angry vectors veer
Round your sleeping head, and form.
There's never need to fear
Violence of the poor world's abstract storm.

Lullaby in *Encounter* May 1957

James Watt 1736–1819

6   James Watt, Who, directing the force of an original genius, Early exercised in philosophic research to the improvement of THE STEAM ENGINE, enlarged the resources of his country, increased the power of men, and rose

to an eminent place among the most illustrious followers of science and
the real benefactors of the world.
Epitaph in Westminster Abbey, London

Warren Weaver 1894–

1   The century of biology upon which we are now well embarked is no matter
of trivialities. It is a movement of really heroic dimensions, one of the great
episodes in man's intellectual history. The scientists who are carrying the
movement forward talk in terms of nucleo-proteins, of ultracentrifuges, of
biochemical genetics, of electrophoresis, of the electron microscope, of
molecular morphology, of radioactive isotopes. But do not be fooled into
thinking this is mere gadgetry. This is the dependable way to seek a
solution of the cancer and polio problems, the problem of rheumatism and
of the heart. This is the knowledge on which we must base our solution of
the population and food problems. This is the understanding of life.
Letter to H M H Carson in *The Story of the Rockefeller Foundation* 1952

Simone Weil 1909–1943

2   *La science, aujourd'hui, cherchera une source d'inspiration audessus d'elle ou
périra.*
*La science ne présente que trois intérêts: 1. les applications techniques;
2. jeu d'échecs; 3. chemin vers Dieu. (Le jeu d'échecs est agrémenté de
concours, prix et medailles.)*
Science today must search for a source of inspiration higher than itself or it
must perish.
Science offers only three points of interest: 1. technical applications;
2. as a game of chess; 3. as a way to God. (The chess-game is embellished
with competitions, prizes and medals.)
La Pesanteur et la Grace 1967 (Paris: Librairie Plon)

Alvin M Weinberg 1915–

3   I would therefore sharpen the criterion of scientific merit by proposing
that, other things being equal, that field has the most merit which contri-
butes most heavily to, and illuminates most brightly, its neighbouring
scientific disciplines.
Minerva 1963 **2** 159–171

Paul Alfred Weiss 1898–

4   A system . . . is exactly the opposite of a machine, in which the structure
of the product depends crucially on strictly predefined operations of the
parts. In the system, the structure of the whole determines the operation of
the parts; in the machine, the operation of the parts determines the outcome.
Beyond Reductionism ed A Koestler and V R Smithies, 1968 (London: Hutchinson)

Carl Friedrich von Weissacker 1912–

5   Those reductionists who try to reduce life to physics usually try to reduce
it to primitive physics—not to good physics. Good physics is broad enough
to contain life, to encompass life in its description since good physics allows

a vast field of possible descriptions. There is no reason why living beings should be compared to primitive machines which don't make use of feedback.

Theoria to Theory 1968, vol 3

Victor Frederick Weisskopf 1908–

1   Science has become adult; I am not sure whether scientists have.

Scientists in Search of their Conscience ed A R Michaelis and H Harvey

2   The value of fundamental research does not lie only in the ideas it produces. There is more to it. It affects the whole intellectual life of a nation by determining its way of thinking and the standards by which actions and intellectual production are judged. If science is highly regarded and if the importance of being concerned with the most up-to-date problems of fundamental research is recognized, then a spiritual climate is created which influences the other activities. An atmosphere of creativity is established which penetrates every cultural frontier. Applied sciences and technology are forced to adjust themselves to the highest intellectual standards which are developed in the basic sciences. This influence works in many ways: some fundamental students go into industry; the techniques which are applied to meet the stringent requirements of fundamental research serve to create new technological methods. The style, the scale, and the level of scientific and technical work are determined in pure research; that is what attracts productive people and what brings scientists to those countries where science is at the highest level. Fundamental research sets the standards of modern scientific thought; it creates the intellectual climate in

which our modern civilization flourishes. It pumps the lifeblood of idea
and inventiveness not only into the technological laboratories and factories,
but into every cultural activity of our time. The case for generous support
for pure and fundamental science is as simple as that.

Why pure science? in the *Bulletin of the Atomic Scientists* 1965 **21** 4–8

Chaim Weizmann 1874–1952

1   [To Meyer Weisgal, Director of the Weizmann Institute] Never let the
scientists get near the *Shissel* [container, where the money is kept].

in Joseph Wechsberg *A Walk through the Garden of Science* 1967 (London: Weidenfeld &
Nicolson)

Arthur Mellen Wellington 1847–1895

2   Engineering . . . is the art of doing that well with one dollar, which any
bungler can do with two after a fashion.

The Economic Theory of the Location of Railways 6th edn, 1900 (New York: Wiley)

Herbert George Wells 1866–1946

3   . . . my epitaph. That, when the time comes, will manifestly have to be: 'I
told you so. You *damned* fools.' (The italics are mine.)

Preface to the 1941 edition of *The War in the Air* originally written in 1907

4   I must confess that I believe quite firmly that an inductive knowledge of a
great number of things in the future is becoming a human possibility. So
far nothing has been attempted, so far no first-class mind has ever focused
itself upon these issues. But suppose the laws of social and political development, for example, were given as many brains, were given as much attention,
criticism and discussion as we have given to the laws of chemical composition during the last fifty years—what might we not expect?

The Discovery of the Future Lecture at the Royal Institution, 1902

5   In England we have come to rely upon a comfortable time-lag of fifty
years or a century intervening between the perception that something ought
to be done and a serious attempt to do it.

The Work, Wealth and Happiness of Mankind 1934 (London: Heinemann) ch 2

6   Queen Victoria was like a great paper-weight that for half a century sat
upon men's minds, and when she was removed their ideas began to blow
about all over the place haphazardly.

The Time Traveller 1973 (London: Weidenfeld & Nicolson)

7   There comes a moment in the day, when you have written your pages in
the morning, attended to your correspondence in the afternoon, and have
nothing further to do. Then comes that hour when you are bored; that's
the time for sex.

in N and J Mackenzie *H G Wells* 1973 (New York: Simon & Schuster)

Rebecca West 1892–

8   Before a war military science seems a real science, like astronomy; but
after a war it seems more like astrology.

Hermann Weyl 1885–1955

1   The whole is always more, is capable of a much greater variety of wave states, than the combination of its parts . . . . In this very radical sense, quantum physics supports the doctrine that the whole is more than the combination of its parts.
Philosophy of Mathematics and Natural Science 1949 (Princeton, NJ: Princeton UP)

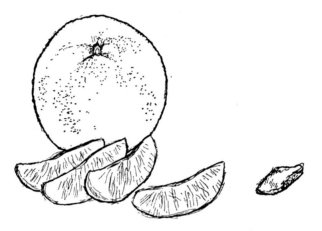

John Archibald Wheeler 1911–

2   There is nothing in the world except empty curved space. Matter, charge, electromagnetism and other fields are only manifestations of the curvature of space.
1957. In *New Scientist* 26 September 1974

3   Time is defined so that motion looks simple.
Gravitation 1973 (Reading: Freeman)

William Whewell 1794–1866

4   As we read the *Principia* [of Newton] we feel as when we are in an ancient armoury where the weapons are of gigantic size; and as we look at them we marvel what manner of man was he who could use as a weapon what we can scarcely lift as a burden.
in E N da C Andrade *Newton and the Science of his Age. Proceedings of the Royal Society* 6 May 1943

5   We need very much a name to describe a cultivator of science in general. I should incline to call him a scientist.
[The first use of the word]
The Philosophy of the Inductive Sciences 1840

James Abbott McNeil Whistler 1834–1903

6   [Answering Oscar Wilde's 'I wish I had said that'] You will, Oscar, you will.
in L C Ingleby *Oscar Wilde*

Alfred North Whitehead 1861–1947

1   The aims of scientific thought are to see the general in the particular and the eternal in the transitory.

2   A crystal lacks rhythm from excess of pattern, while a fog is unrhythmic in that it exhibits a patternless confusion of detail.

3   It is a profoundly erroneous truism, repeated by all copy books and by eminent people when they are making speeches, that we should cultivate the habit of thinking of what we are doing. The precise opposite is the case. Civilization advances by extending the number of important operations which we can perform without thinking about them.

4   It is a safe rule to apply that, when a mathematical or philosophical author writes with a misty profundity, he is talking nonsense.
An Introduction to Mathematics 1948 (Oxford: Oxford UP)

5   The possibilities of modern technology were first in practice realised in England by the energy of a prosperous middle class. Accordingly, the industrial revolution started there. But the Germans explicitly realised the methods by which the deeper veins in the mine of science could be reached. In their technological schools and universities progress did not have to wait for the occasional genius or the occasional lucky thought. Their feats of scholarship during the nineteenth century were the admiration of the world. This discipline of knowledge applies beyond technology to pure science, and beyond science to general scholarship. It represents the change from amateurs to professionals.
Science and the Modern World 1926 (London: Cambridge UP)

6   Science is taking on a new aspect which is neither purely physical nor purely biological. It is becoming the study of organisms. Biology is the study of the larger organisms; whereas physics is the study of the smaller organisms.
Science and the Modern World 1926 (London: Cambridge UP)

Walt Whitman *ca* 1819–1892

7   Do I contradict myself? Very well then I contradict myself (I am large, I contain multitudes).
Song of Myself 1938 (London: Nonesuch Press) 5

8   I need no assurances . . .
I do not doubt that temporary affairs keep on and on millions of years.
I do not doubt that interiors have their interiors and exteriors have their exteriors . . .
Assurances in *Nonesuch Edition of Collected Poems* 1938 (London: Nonesuch Press)

9   When I heard the learn'd astronomer,
When the proofs, the figures, were ranged in columns before me,

When I was shown the charts and diagrams, to add, divide, and measure them,
When I sitting heard the astronomer where he lectured with much applause in the lecture-room,
How soon unaccountable I became tired and sick,
Till rising and gliding out I wander'd off by myself,
In the mystical moist night-air, and from time to time,
Look'd up in perfect silence at the stars.

[1865–1867]
When I heard the learn'd astronomer in Nonesuch Edition of Collected Poems 1938 (London: Nonesuch Press)

Benjamin Lee Whorf 1897–1934

1   . . . an explicit scientific world view may arise by a higher specialization of the same basic grammatical patterns that fathered the naive and implicit view. Thus the world view of modern science arises by higher specialization of the basic.

Language, Thought and Reality 1956 (Cambridge, Mass: The MIT Press)

Norbert Wiener 1894–1964

2   A painter like Picasso, who runs through many periods and phases, ends up by saying all those things which are on the tips of the tongues of the age to say, and finally sterilises the originality of his contemporaries and juniors.

The Human Use of Human Beings 1950 (London: Sphere Books)

3   We are raising a generation of young men who will not look at any scientific project which does not have millions of dollars invested in it . . . . We are for the first time finding a scientific career well paid and attractive to a large number of our best young go-getters. The trouble is that scientific work of the first quality is seldom done by the go-getters, and that the dilution of the intellectual milieu makes it progressively harder for the individual worker with any ideas to get a hearing . . . . The degradation of the position of the scientist as an independent worker and thinker to that of a morally irresponsible stooge in a science-factory has proceeded even more rapidly and devastatingly than I had expected.

Bulletin of the Atomic Scientists 4 November 1948, pp338–9

Jerome Bert Wiesner 1915–

4   Some problems are just too complicated for rational logical solutions. They admit of insights, not answers.

in D Lang *Profiles: A Scientist's Advice II. New Yorker* 26 January 1963

Eugene Paul Wigner 1902–

5   The simplicities of natural laws arise through the complexities of the languages we use for their expression.

Communications on Pure and Applied Mathematics 1959 **13** 1

Oscar Wilde 1854–1900

6   Experience is the name everyone gives to their mistakes.

Lady Windermere's Fan

1   Religions die when they are proved to be true. Science is the record of dead religions.
Phrases and Philosophies for the Use of the Young 1891

Helmuth Wilhelm 1905–

2   Change, that is the only thing in the Universe which is Unchanging.
[Epitomizing the *I Ching*, 8th Century BC]
Der Zeitbegriff im Buch der Wandlungen in *Eranos Jahrbuch* 1951 **20** 321

William Henry [Duke of Gloucester] 1743–1805

3   Another damned, thick, square book. Always scribble, scribble, scribble. Eh: Mr Gibbon?
Quoted in a Note to Boswell's *Life of Johnson*

William of Occam 1300–1349

4   [Occam's Razor] *Entia non sunt multiplicanda praeter necessitatem.*
It is vain to do with more what can be done with less.
Attributed

[Lord] John [of Selmeston] Wilmot 1895–1964

5   What I like about scientists is that they are a team, so that one need not know their names.
[Minister of Supply 1945–1947]
The Prof in two Worlds 1961 (London: Collins)

[Sir] Harold Wilson 1916–

6   If there was one word I could use to identify modern socialism it was 'science'.
[But the 'white-hot technological revolution' never took place and, like Christianity, science was never really tried]
The Relevance of British Socialism 1964 (London: Weidenfeld & Nicolson)

Wallace Wilson 20th Century

7   He prayeth best who loveth best
All creatures great and small.
The Streptococcus is the test
I love him least of all.
[In parody of Coleridge's *Ancient Mariner*]

Ludwig Wittgenstein 1889–1951

8   In order to draw a limit to thinking, we should have to think both sides of this limit.
Tractatus Logico-Philosophicus 1961 (London: Routledge & Kegan Paul)

9   We could present spatially an atomic fact which contradicted the laws of physics, but not one which contradicted the laws of geometry.
Tractatus Logico-Philosophicus 1961 (London: Routledge & Kegan Paul)

10   We feel that even if all possible scientific questions be answered, the problems of life have still not been touched at all. Of course there is then no

question left, and just this is the answer. The solution of the problem of life is seen in the vanishing of this problem.

Tractatus Logico-Philosophicus 1961 (London: Routledge & Kegan Paul)

1   Whereof one cannot speak, thereof one must be silent.

Tractatus Logico-Philosophicus 1961 (London: Routledge & Kegan Paul)

Friedrich Woehler 1800–1882

2   Organic chemistry just now is enough to drive one mad. It gives one the impression of a primeval, tropical forest full of the most remarkable things, a monstrous and boundless thicket, with no way of escape, into which one may well dread to enter.

Letter to Berzelius 28 January 1885

[Cardinal] Thomas Wolsey *ca* 1475–1530

3   This new invention of printing has produced various effects of which Your Holiness cannot be ignorant. If it has restored books and learning, it has also been the occasion of those sects and schisms which daily appear. Men begin to call in question the present faith and tenets of the Church; the laity read the scriptures and pray in their vulgar tongue. Were this suffered the common people might come to believe that there was not so much use of the clergy. If men were persuaded that they could make their own way to God, and in their ordinary language as well as Latin, the authority of the Mass would fall, which would be prejudicious to our ecclesiastical orders. The mysteries of religion must be kept in the hands of the priests.

Frederic Wood-Jones 1879–1954

4   Whoever wins to a great scientific truth will find a poet before him in the quest.

Medical Journal of Australia 29 August 1931

[Sir] Richard van der Riet Woolley 1906–

5   Space-travel is utter bilge.

[The Astronomer-Royal, 1956–1971. Quoted *ca* 1956]
in Arthur C Clarke *Profiles of the Future* 1973 (London: Gollancz)

Elizabeth Wordsworth 1840–1932

6   If all the good people were clever;
And all clever people were good,
The world would be nicer than ever
We thought that it possibly could.

Good and Clever

William Wordsworth 1770–1850

7   . . . . Where the statue stood
Of Newton with his prism and silent face,
The marble index of a mind for ever
Voyaging through strange seas of Thought, alone.

[Newton's Statue at Trinity College, Cambridge]
The Prelude 1850, Book III, written 1779–1805

1　　Lost in a gloom of uninspired research.
　　　The Excursion Book 4

2　　Man now presides
　　　In power, where once he trembled in his weakness;
　　　Science advances with gigantic strides;
　　　But are we aught enriched in love and meekness?
　　　To the Planet Venus 1838

3　　Physician art thou—one, all eyes,
　　　Philosopher—a fingering slave,
　　　One that would peep and botanize
　　　Upon his mother's grave.
　　　A Poet's Epitaph

4　　Poetry is the breath and finer spirit of all knowledge; it is the impassioned
　　　expression which is in the countenance of all Science . . . shall be ready to
　　　put on, as it were, a form of flesh and blood, the poet will lend his divine
　　　spirit to aid the transfiguration, and will welcome the Being thus produced
　　　as a dear and genuine inmate of the household of man.

5　　To the solid ground of nature trusts the Mind that builds for aye.
　　　Quotation appearing on the title page of *Nature* until 1963

6　　Yet we may not entirely overlook
　　　The pleasure gathered from the rudiments
　　　Of geometric science . . .
　　　The Prelude 1850, Book VI, lines 115–7

John Wycliffe *ca* 1320–1384
7　　God forceth not a man to believe that which he cannot understand.
　　　[Translator of the Bible into English]

Xenophon *ca* 444–*ca* 354 BC
8　　What are called the mechanical arts carry a social stigma and are rightly
　　　dishonoured in our cities. For these arts damage the bodies of those who
　　　work at them or who act as overseers, by compelling them to a sedentary
　　　life and, in some cases, to spend the whole day by the fire. This physical
　　　degeneration results also in deterioration of the soul. Furthermore, the
　　　workers in these trades simply have not got the time to perform the offices
　　　of friendship or citizenship. Consequently they are looked on as bad friends
　　　and bad patriots, and in some cities, especially the warlike ones, it is not
　　　legal for a citizen to ply a mechanical trade.
　　　Oeconomicus

Yang Hsiung 51 BC–AD 18
9　　Someone asked whether a sage could make divination. [Yang Hsiung]

replied that a sage could certainly make divination about Heaven and Earth. If that is so, continued the questioner, what is the difference between the sage and the astrologer (*shih*)? [Yang Hsiung] replied, 'The astrologer foretells what the effects of heavenly phenomena will be on man; the sage foretells what the effects of man's actions will be on the heavens'.

Fa Yen (Model Discourses) *ca* 5 AD. Transl J Needham *Science and Civilization in China* 1956 (London: Cambridge UP)

William Butler Yeats 1865–1939

1   The intellect of man is forced to choose
Perfection of the life, or of the work.
The Choice in *The Collected Poems of W B Yeats* 1933 (London: Macmillan)

2   Locke sank into a swoon;
The Garden died;
God took the spinning-jenny
Out of his side.
Fragments in *The Collected Poems of W B Yeats* 1933 (London: Macmillan)

3   Science is the religion of the suburbs.
The Spectator 5 July 1969

Robert M Yerkes 1876–1956

4   One chimpanzee is not a chimpanzee at all.
Chimpanzees, a Laboratory Colony 1945 (New Haven, Conn: Yale UP)

Edward Young 1683–1765

5   Love finds admission, where proud science fails.
Night Thoughts 1742–1745

6   This gorgeous apparatus
This display
This ostentation of creative power
This theatre
What eye can take it in . . .?
The Complaint, or Night Thoughts, Night Ninth 1744

Yuan Mei 1716–1797

7   When the lists go up much is heard of the candidates' resentment; No one realizes with what sadness the examiners did their duty.
Sir Douglas Logan in the *University of London Bulletin*. See also *Yuan Mei* 1956 (London: Allen & Unwin)

Yevgeni Ivanovich Zamyatin 1884–1937

8   Tell me, what is the final integer, the one at the very top, the biggest of all? But that's ridiculous! Since the number of integers is infinite, how can you have a final integer?
Well then how can you have a final revolution?

There is no final revolution. Revolutions are infinite.

We transl W N Vickery in P Blake and M Hayward *Dissonant Voices in Soviet Literature* 1962 (New York: Random House)

Index

Acknowledgments

Grateful acknowledgment is made to the following for their kind permission to reprint copyright material. Every effort has been made to trace copyright ownership but if, inadvertently, any mistake or omission has occurred, full apologies are herewith tendered.

Full references to authors, the titles of their works, and publishers are given under the appropriate quotation.

Addison–Wesley Publishing Co, London

Akademische Verlagsgesellschaft, Leipzig

George Allen & Unwin Ltd, Hemel Hempstead

American Association of Physics Teachers, New York

American Institute of Physics, New York

American Physical Society, New York

Architectural Press Ltd, London

Associated Book Publishers Ltd, London

Association of University Teachers, London

Basic Books Inc, New York

Baskervilles Investments Ltd, London

Professor Stafford Beer

G Bell & Sons Ltd, London

Adam & Charles Black Ltd, London

The Bodley Head, London

British Association for the Advancement of Science, London

British Broadcasting Corporation (Publications), London

Rita Bronowski for the Estate of Jacob Bronowski

The Bulletin of the Atomic Scientists, Chicago, Illinois

John Calder (Publishers) Ltd, London

Cambridge University Press, London

Campbell Thomson & McLaughlin Ltd, London

Jonathan Cape Ltd, London

Cassell & Collier Macmillan Publishers Ltd, London

Chatto & Windus Ltd, London

The Ciba Foundation, London

Ronald W Clark

Arthur C Clarke

Miss D E Collins

William Collins, Sons & Co Ltd, London

Committee of Vice-Chancellors and Principals of the Universities of the United Kingdom, London

Constable & Co, Ltd, London

Curtis Brown Ltd, London

J M Dent & Sons Ltd, London

Andre Deutsch Ltd, London

Diogenes Paris

Elizabeth H Dos Passos

Doubleday & Company Inc, New York

Dover Publications Inc, New York

The Economist Newspaper Limited, London

Edinburgh University Press, Edinburgh

Edition Stock, Paris

Editions Gallimard, Paris

Elek Books Ltd, London

Elsevier Publishing Co, Barking

Encounter Encounter Ltd, London

Encyclopaedia Britannica Inc, Chicago, Illinois

D J Enright

Estate of Mrs George Bambridge

Estate of E C Bentley

Estate of Albert Einstein

Estate of Robert Frost

Estate of C Day Lewis

Estate of George Orwell

Estate of Bertrand Russell

Estate of Lord Tweedsmuir

Estate of H G Wells

Evening Standard Beaverbrook Newspapers Ltd, London

Faber & Faber Ltd, London

R Buckminster Fuller

Gaberbocchus Press Ltd, London

Victor Gollancz Ltd, London

Professor Samuel A Goudsmit

Robert Graves

Warren H Green Inc, St Louis, Missouri

The Guardian London

Mrs Helen Spurway Haldane

Hamish Hamilton Ltd, London

The Hamlyn Publishing Group Limited, Feltham

Harcourt Brace Jovanovich, Inc, New York

Harper & Row, Publishers, Inc, New York

Thomas G Hart

Harvard University Press, Cambridge, Massachusetts

A M Heath & Co Ltd, London

William Heinemann Ltd, London

Heinemann Educational Books Ltd, London

Her Majesty's Stationery Office, London

David Higham Associates Ltd, London

History of Science Society, Washington, DC

The Hogarth Press Ltd, London

Holt, Rinehart and Winston Inc, New York

Houghton Mifflin Co, Boston, Massachusetts

Humanities Press Inc, Atlantic Highlands, NJ

Hutchinson Publishing Group Ltd, London

Mrs Laura Huxley

IEEE Spectrum New York

Institute of Theoretical Physics, Calcutta

International Statistical Institute, Voorburg

The International Union of Biological Sciences, Paris

The Johns Hopkins University Press, Baltimore, Maryland

Mrs Katherine Jones

Professor R V Jones

Alfred A Knopf Inc (a division of Random House Inc), New York

The Lancet London

Lawrence & Wishart Ltd, London

Tom Lehrer

Claude Levi-Strauss

Librairie Hachette, Paris

Librairie Plon, Paris

Arthur D Little Inc, Cambridge, Massachusetts

Little, Brown & Co, Boston, Massachusetts

Macmillan & Co Ltd, London and Basingstoke

Macmillan Publishing Co Inc, New York

Sir Peter Medawar

The Medical Journal of Australia Glebe, NSW

The Merlin Press Ltd, London

Julian Messner Inc (a division of Simon & Schuster Inc), New York

Minerva London

The MIT Press, Cambridge, Massachusetts

William Morrow & Co Inc, New York

John Murray (Publishers) Ltd, London

National Aeronautics and Space Administration, Washington, DC

Nature Macmillan (Journals) Ltd, London

Dr J Needham

The New American Library Inc, New York

New Directions Publishing Co, New York

New Scientist (the weekly review of science and technology) New Science Publications, London

The New York Academy of Sciences, New York

The New York Review of Books Copyright © 1970 Nyrev, Inc, New York

New York University Press, New York

The New Yorker New York

North-Holland Publishing Company, Amsterdam

W W Norton & Company Inc, New York

Harold Ober Associates Inc, New York

The Observer London

Orell Füssli Verlag, Zurich

Mrs Sonia Brownell Orwell

Peter Owen Ltd, London

Oxford University Press, New York

Oxford University Press, Oxford

Pantheon Books Inc (a division of Random House Inc), New York

C Northcote Parkinson

Pemberton Publishing Co Ltd, London

Penguin Books Ltd, London and Harmondsworth

Pergamon Press Ltd, Oxford

A D Peters & Co, London

Phaidon Press Ltd, London

The Physical Review Long Island, NY

Physical Review Letters Long Island, NY

Professor Sir Brian Pippard

Pitman Publishing Ltd, London

Praeger Publishers Inc, New York

Presses Universitaires de France, Paris

Princeton University Press, Princeton, NJ

G P Putnam's Sons, New York

Random House Inc, New York

D Reidel Publishing Company, Dordrecht

Revista de Occidente SA, Madrid

Kenneth Rose

Routledge & Kegan Paul Ltd, Henley-on-Thames

Royal Geographical Society, London

The Royal Institution, London

The Royal Society, London

Salk Institute, San Diego, California

Sather Classical Lectures Copyright © 1952 by the Regents of the University of California

Saturday Review New York

John Schaffner Literary Agency, New York

Schenkman Publishing Company, Cambridge, Massachusetts

Science American Association for the Advancement of Science, Washington, DC

Science Policy Foundation, London

Scientific American W H Freeman & Co, San Francisco, California

SCM Press Ltd, London

The Scotsman Edinburgh

Charles Scribner's Sons, New York

Martin Secker & Warburg Ltd, London

Sheed & Ward Ltd, London

Simon & Schuster Inc, New York

Martyn Skinner

The Society of Authors for the Estates of Gordon Bottomley, John Masefield and Bernard Shaw

The Society of the Friends of Chester Cathedral

Souvenir Press Ltd, London

The Spectator London

Spiegel-Verlag, Hamburg

Springer-Verlag, New York

Stanford University Press, Stanford, California

Sunday Telegraph London

Stefan Themerson

Charles C Thomas, Publisher, Springfield, Illinois

H W Tilman

Time Magazine The Weekly News Magazine © Time Inc 1976

The Times Times Newspapers Ltd, London

UNESCO, Paris

University of California Press, Berkeley, California

The University of Chicago Press, Chicago, Illinois

University of London Bulletin London

University of Pennsylvania Press, Philadelphia, Pennsylvania

University of Toronto Press, Toronto

Mrs M J Waddington for the Estate of Professor C H Waddington

Walker & Co, New York

A P Watt & Son, London

C A Watts & Co Ltd, London

Weidenfeld & Nicolson Ltd, London

John Wiley & Sons Inc, New York

John Wiley & Sons Ltd, Chichester

World Federation of Scientific Workers, London

Miss Anne Yeats

M B Yeats

Znanie, Moscow